Lecture Notes in Morphogenesis

Lecture Notes in Morphogenesis is an interdisciplinary book series which aims to face the questions of emergence, individuation and becoming of forms from several different points of view: those of pure and applied mathematics, of computational algorithms, of biology, of neurophysiology, of cognitive and social structures. The set of questions above concerns all the manifestations of Being, all the manifestations of Life. At the heart of contemporary embryogenesis lies an essential question: How can form emerge from the constant, chaotic flow? How can a sequence of purely informational elements — an a-signifying combination of chemical substances organized in the DNA molecule — evolve into the highly complex and structured forms of the living organism? A similar question can be asked when we deal with the morphogenesis of vision in neural systems and with the creation of evolving synthetic images, since digital technology makes possible the simulation of emergent processes both of living bodies and of visual forms. Finally the very idea that abstract structures of meaning could be captured in terms of morphodynamic evolution opens the door to new models of semiolinguistics, semiotic morphodynamics, and cognitive grammars. An entire heritage of ideas and concepts has to be reconsidered in order to face new and challenging problems: the theoretical framework opened by Goethe with the introduction of the word "Morphogenesis" is developed by D'Arcy Thompson in "On Growth and Form", it is reorganized with new theoretical insights by the classical structuralism of Levi-Strauss and formalized by the dynamical structuralism of René Thom. The introduction of the post-structuralists ideas of individuation (in Gilbert Simondon and Gilles Deleuze) and plasticity of structures builds a bridge to contemporary problems of morphogenesis at a physical, biological, social and transindividual level. The objective of this book series is to provide suitable theoretical and practical tools for describing evolutionary phenomena at the level of Free boundary problems in Mathematics, Embryogenesis, Image Evolution in Visual Perception, Visual Models of Morphogenesis, Neuromathematics, Autonomy and Self-Organization, Morphogenetic Emergence and Individuation, Theoretical Biology, Cognitive Morphodynamics, Cities Evolution, Semiotics, Subjectivation processes, Social movements as well as new frontiers of Aesthetics. To submit a proposal or request further information, please use the PDF Proposal Form or contact directly: *Dr. Thomas Ditzinger (thomas.ditzinger@springer.com)*

Luca Taddio

The Appearing of Sense

From Gestalt Theory to Merleau-Ponty Phenomenology

Springer

Luca Taddio
DIUM
Udine University
Udine, Italy

ISSN 2195-1934 ISSN 2195-1942 (electronic)
Lecture Notes in Morphogenesis
ISBN 978-3-032-27020-7 ISBN 978-3-032-27021-4 (eBook)
https://doi.org/10.1007/978-3-032-27021-4

This Springer imprint is published by the registered company Springer Nature Switzerland AG
The registered company address is: Gewerbestrasse 11, 6330 Cham, Switzerland

Only as an aesthetic phenomenon is existence and the world eternally justified.

—F. Nietzsche

Competing Interests The author has no competing interests to declare that are relevant to the content of this manuscript.

Preliminary Remarks

Starting from a critical reconsideration of Merleau-Ponty's relation to Gestalt psychology, this study aims to explore the preliminary sphere of understanding in which meaning first emerges. From a phenomenological standpoint, our intent is twofold: first, to show that *appearing as such* is not a matter of falsifiability (whereas judgments may be true or false, appearing itself is not) and second, to argue that the relation between appearance and reality must be conceived as epistemological rather than ontological. To pursue this line of inquiry, it becomes necessary to clarify the meaning of "world," distinguishing it both from "reality" and from what we call "the real."

Merleau-Ponty brings to the fore a *wild world*, which, for us, embodies the very virtuality of being. To enter into his work is to conceive the world as the virtual dimension of reality: to rethink the nexus between appearance and reality, to interpret illusion in an ontological sense, and more broadly, to understand the entire field of appearing as a non-falsifiable dimension that precedes all judgment and epistemological positioning. What we aim to outline here is a phenomenology that may be termed *heretical*: one that unfolds and branches out, diverging from Husserl while remaining deeply rooted in a vast and partly autonomous psychological tradition.

These are complex intertwining, whose legacy is not easily assumed but whose theoretical provocation demands to be taken up anew. Our working hypothesis thus situates Merleau-Ponty at a decisive crossroads, one at which paths diverge or converge depending on the perspective adopted: a dual movement that, on the one hand, radiates outward toward unexplored directions and, on the other, converges upon a fundamental core of thought. This movement culminates in a philosophy of relation, which we attempt to trace through a series of reflections accompanying a phenomenology of perception, *the* foundation and theoretical epicenter of Merleau-Ponty's philosophy. Our aim is to rediscover his thought as a space in which the strings of phenomenology may resonate in a renewed way. Entering his work thus means seeking out its metaphysical core, discerning that "ray of world" capable of guiding us toward phenomenology as an original stance within the *Lebenswelt*.

Philosophy as Merleau-Ponty reminds us, must interrogate its own presuppositions: unlike other forms of knowledge, it can accept no premise unexamined, but must continuously problematize its point of departure. To philosophize *through* and *with* Merleau-Ponty is first of all to elucidate the immediacy of experience and to clarify the mode by which we relate to other domains of knowledge. To begin with perception is to recognize it as an opening to the truth of being, an opening that requires us to thematize our access to the world through an analysis of corporeality. To understand the origin of knowledge is each time to assume a new situation, one that compels us to reopen phenomenological inquiry within a horizon of meaning that demands to be both grasped and comprehended. While we perceive a world that appears relatively stable, we inhabit a reality perpetually in transformation. The intertwining of these two dimensions calls for an analysis of the relation between the visible and the invisible, and at once a critical examination of the points at which epistemology and ontology intersect or diverge.

The meaning of phenomena unfolds within a field of existence that our body institutes in relation to its surrounding environment. From this primordial relation both the world and subjectivity emerge. Appearance thus takes shape as a virtuality internal to reality itself. This perspective allows for a reading of Merleau-Ponty that can enter into dialogue with thinkers seemingly distant from phenomenology in the strict sense, such as Gilbert Simondon and Gilles Deleuze.

The concept of truth that arises from Merleau-Ponty's analysis of perception must be understood as an *opening*—that is, as a relation—to the world. It recalls the original sense of *aletheia*: unveiling, the disclosure of Being. The manifestation of phenomena constituting the visible must display a degree of stability—never absolute nor fixed, but marked by a certain *meta-stability* and coherence—without which no access to reality would be possible. Were the world incoherent or reduced to mere representation, any aspiration to knowledge would dissolve.

What heuristic value could the natural sciences possess if the phenomenal field were radically contradictory? The world, however, is not an image nor a mirror. We do not inhabit the closed frame of Magritte's *The Human Condition*, with its painting-within-a-painting superimposed upon the landscape. Such a metaphor misleads, for we are always incarnated in the world: there is neither a subject without world nor a world without a subjectivity that lives and experiences it. There is one reality, yet as many worlds as there are forms of life. The lived body and the world are two faces of a single dynamic structure, a *Gestalt* in which each implies and sustains the other. The world is the correlate of experience, revealing a plane of reality composed of folds, levels, and internal relations within Being. It is in this sense that Merleau-Ponty's monism should be understood. The world may be conceived as one level of reality, from which, on empirical-phenomenological grounds, we can inquire, albeit indirectly through the empirical-experimental sciences, into the invisible strata of Being.

Modern science, by multiplying domains and continually redrawing their borders, tends to constrict the domain of philosophy. Yet even as this contraction occurs, an irreducible remainder persists between science and philosophy, a residue that legitimises the need for phenomenological reflection. It is thus necessary to define certain theoretical traits capable of restoring an idea of philosophy grounded in the methodological premises of the phenomenology of perception.

Phenomenology can contribute to epistemological clarification not only through systematic use of real experiments or *Gendankenexperiments*, but also through the analysis of examples drawn from scientific practice. In this light, Merleau-Ponty's oeuvre can be read, in a certain sense, as an extended critical footnote to both Husserl and Gestalt psychology. He reinterprets Husserl through Gestalt theory while, conversely, using Husserlian insights to question the "new psychology" that transcends the limits of *Gestaltpsychologie*.

Far from diminishing his originality, this double movement—simultaneously interpretative and inventive—defines Merleau-Ponty's method and configures his thought as strikingly contemporary. It also marks the direction in which his legacy may still unfold. Our inquiry focuses particularly on the relation between Merleau-Ponty and the psychology of perception, where the possibility emerges for a renewal not only of his thought but of phenomenology itself. The aim is to make visible the fundamental concepts of his philosophy by embodying them through concrete cases, primarily drawn from his own writings. These examples serve not merely illustrative but paradigmatic functions: singularities through which epistemological and ontological implications may be discerned. Although selective and interpretative, this reading engages the theoretical and methodological tensions at the very foundation of phenomenology.

We might therefore apply to Merleau-Ponty what he once said of Husserl: a philosopher's thought cannot be reduced to analytical observation or isolating reflection, for such an approach would obliterate its finest nuances. It must instead be recovered through "thinking again" (S, 160). To approach Merleau-Ponty's work in purely analytical terms would fragment his philosophy into discrete elements, stripping it of its living coherence. By contrast, to reactivate our original contact with the world, to embrace it in its full complexity, is to recover the movement animating his work: to understand the figure of the author against the visible–invisible background that sustains his reflections. Beginning from perception, our task is to traverse the guiding lines of Merleau-Ponty's thought in order to reach the core of his philosophy.

Abbreviations of Merleau-Ponty's Works

C *Causeries 1948*, texte annoté par S. Ménasé, Paris: Éditions du Seuil, 2002.

EM *L'OEil et l'esprit*. Paris: Gallimard, 1964 (folio essais). /*"Eye and Mind." The Primacy of Perception and Other Essays on Phenomenological Psychology, the Philosophy of Art, History and Politics*. J. M. Edie, ed. Evanston: Northwestern University Press, 1964.

EP *Éloge de la philosophie*. Paris: Gallimard, 1953. *Éloge de la philosophie et autres essais. Paris: Gallimard, 1960 (folio essais)./In Praise of Philosophy and Other Essays*. J. Wild, J. Edie, and J. O'Neill, trans. Evanston: Northwestern University Press, 1988.

N *La Nature: Notes, Cours du Collège de France. D. Séglard, ed. Paris: Seuil, 1994./Nature*. D. Séglard, ed. R.Vallier, trans. Evanston: Northwestern University Press, 2003.

PP *Phénoménologie de la perception. Paris: Gallimard, 1945./Phenomenology of Perception*. C. Smith, trans. London: Routledge & Kegan Paul, 1962./London and New York: Routledge, 2002.

PrP *Le Primat de la perception et ses conséquences philosophiques*. Verdier, 1996. *The Primacy of Perception and Other Essays on Phenomenological Psychology, the Philosophy of Art, History and Politics*. J. M. Edie, ed. Evanston: Northwestern University Press, 1964.

RC *Résumés de cours, Collège de France 1952–1960. Paris: Gallimard, 1968./ "Themes from the Lectures at the Collège de France, 1952– 1960." In Praise of Philosophy and Other Essays*. J. Wild, J. Edie, and J. O'Neill, trans. Evanston: Northwestern University Press, 1988.

S *Signes*. Paris: Gallimard, 1960./*Signs*. R. McCleary, trans. Evanston: Northwestern University Press, 1964.

SB *La Structure du comportement. Paris: Presses Universitaires de France, 1942. The Structure of Behavior*. A. Fisher, trans. Boston: Beacon Press, 1963.

SNS *Sens et non-sens. Paris: Nagel, 1948./Sense and Non-Sense*. H. L. Dreyfus and P. Dreyfus, trans. Evanston: Northwestern University Press, 1964.

SW *Le monde sensible et le monde de l'expression: Cours au Collège de France Notes* (1953). Genève: Mētis Presses, 2011. *The Sensible World and the World of Expression: Course Notes from the Collège de France, 1953.* B. Smith, trans. Evanston: Northwestern University Pres

VI *Le Visible et l'invisible.* C. Lefort, ed. Paris: Gallimard, 1964. *The Visible and the Invisible.* A. Lingis, trans., Evanston: Northwestern University Press, 1968.

Other Works by Merleau-Ponty

- *Humanisme et terreur, Essai sur le problème communiste.* Paris: Gallimard, 1947./ *Humanism and Terror: An Essay on the Communist Problem.* J. O'Neill, trans. Boston: Beacon Press, 1969.
- *Husserl at the Limits of Phenomenology.* L. Lawlor and B. Bergo, eds. Evanston: Northwestern University Press, 2002.
- *L'Institution dans l'histoire personnelle et publique. Le Problème de la passivité: Le Sommeil, l'inconscient, la m´emoire. Notes de cours au Collège de France* (1954–1955). D. Darmaillacq, C. Lefort, S. Ménasé, eds. Paris: Belin, 2003.
- *L'Union de l'âme et du corps chez Malebranche, Biran et Bergson.* 2d ed. J. Deprun, ed. Paris: Vrin, 2002. First edition pagination. *The Incarnate Subject: Malebranche, Biran, and Bergson on the Union of Body and Soul.* A. G. Bjelland and P. Burke, eds. P. B. Milan, trans. Amherst, N.Y.: Humanities Books, 2001.
- *Les Aventures de la dialectique.* Paris: Gallimard, 1955 *(folio essais). Adventures of the Dialectic.* J. Bien, trans. Evanston: Northwestern University Press, 1973.
- *La Prose du monde.* C. Lefort, ed. Paris: Gallimard, 1969./*The Prose of the World.* J. O'Neill, trans. Evanston: Northwestern University Press, 1973.
- *Notes des Cours au Collège de France: 1958–1959 et 1960–1961.* Paris: Gallimard, 1996.
- *Parcours, 1935–1951. J. Prunair, ed. Paris: Verdier, 1997. Short critical pieces on such figures as Kafka, Levinas, Marcel, Marx, Sartre, and Scheler. Parcours deux, 1951–1961.*
- *Psychologie et pédagogie de l'enfant. Cours de Sorbonne 1949–1952. J. Prunair, ed. Paris: Verdier, 2001.* Includes *"La Conscience et l'acquisition du langage" of 1949–1950*, translated as *Consciousness and the Acquisition of Language.* H. J. Silverman, trans. Evanston: Northwestern University Press, 1973.
- *Texts and Dialogues.* H. J. Silverman and J. Barry, Jr., eds. M. B. Smith, et al., trans. Atlantic Highlands, N. J.: Humanities Press International, 1992.
- *The Merleau-Ponty Aesthetics Reader: Philosophy and Painting.* G. A. Johnson and M. B. Smith, eds. Evanston: Northwestern University Press, 1993.

Contents

Contents

Chapter 1
Introduction

1.1 I

By selecting a few key points and linking them through a series of examples, we aim to clarify the stakes of the problem under discussion. For this reason, we have deliberately minimized the use of specialized vocabulary and technical details. Regarding Merleau-Ponty's biography, we follow Heidegger's remark on Aristotle: that, as far as a philosopher's life is concerned, all that truly matters is that "he was born, worked, and died."[1] We have thus focused on the essential traits of Merleau-Ponty's thought without deepening the episodes of his life. Works of genuine theoretical or artistic significance do not find their meaning in the life of their authors; rather, the life mirrors what those works bring into being. As Merleau-Ponty writes of Cézanne, that "*work required that life*" (SNS, 20). The description of external circumstances, being the contingent outcome of a creative act, can never exhaust the meaning of a work, nor can it be reduced to the analysis of its premises or to the enumeration of its constitutive elements.

An introductory path into Merleau-Ponty's philosophy may likewise serve as an initiation into philosophy itself, on the condition that the encounter with the author becomes a genuine act of reflection: a shared gaze upon the problems and phenomena under consideration, whether experimental or directly lived. Identifying the point of departure, though never neutral or free of theoretical presuppositions, should be safeguarded as far as possible from conceptual prejudice. Why not, then, begin with a critical analysis of that primordial moment of knowledge that is experience itself? Philosophy has indeed followed this path before; yet the empiricist tradition, as we shall see, differs profoundly from phenomenology. While empiricism grounds knowledge on the irreducible parts of things, on sensations rather than perception, phenomenology seeks to restore our original contact with the world, not by disassembling it like a puzzle into discrete elements, but by learning to receive phenomena

[1] Martin Heidegger apparently told it at the beginning of a lecture on Aristotle. Cf. S.J. Mc Grath, *Heidegger: A (Very Critical Introduction*, Grand Rapids: Eerdmans, 2008, p. 8.

L. Taddio, *The Appearing of Sense*, Lecture Notes in Morphogenesis,
https://doi.org/10.1007/978-3-032-27021-4_1

in their living complexity and immediacy. Can this same perspective emerge through art? To grasp the meaning of a work, one cannot fragment it or limit oneself to isolating its generative components. The work must be approached in its expressive immediacy, beginning from the relation that binds us to it. It is precisely in this direction that Merleau-Ponty leads us.

The analysis of perception and its implications thus aims to reach the very beginning of knowledge. If phenomenology is a work of clarification, it must begin from the thematic core out of which the entire path unfolds. Perception "grounds everything, because it shows us, so to speak, an obsessional relation with being" (EP, 16). To return systematically to the thing itself presupposes the idea—and more deeply, the felt awareness—that there is something there which calls for our care, something whose formation does not fully coincide with our will or our timing. Through technique, the artist affirms what demands our work upon matter in order to come into presence: a possible articulation of becoming, which takes shape (*Gestalt*) within a morphogenesis of the sense. Through the forms proper to *poietic* activity, the artist preserves and reveals a sense. This technical operation, though in certain respects analogous to natural processes, is nevertheless commonly opposed to nature, as if some forms required human intervention in order to exist, while others emerged spontaneously. Yet can we not conceive a system of equivalences between a natural landscape and what we perceive in a painting, or rather, through a painting? What is equivalent there, above all, is not substance but an *order of meaning*.

At different scales, *recursive* processes can be observed throughout nature, in both the physical and biological realms. Their dynamic character reveals a dependence on time in the formation of *structures.* Similarly, to meditate, to circle around a problem, or to bring forth an idea (as when we are haunted by a "fixed idea"), entails a continual return to the *thing itself*: cycles, recurrences, and recursive rhythms that accompany generation. These rhythms do not belong to subjectivity alone but recall, rather, the farmer's care for the land, the patient gestures of cultivation that precede the harvest. Every authentic creative act requires its own time of gestation; it cannot be summoned at will, according to the demands of our rhythm or schedule. Art, too, bears the marks of this recursive movement: a work of excavation, research, and meditation. Cézanne's repeated attempts to paint Mont Sainte-Victoire exemplify this process. "A minute of the world is going by," he said, "which must be painted in its full reality" (SNS, 17), in its becoming across the hours of the day, in the instant when a passing cloud casts its shadow upon the mountain. The mountain's becoming and appearing are composed of innumerable colors, shades, shadows, and lights, each containing and generating new forms. Its presence is at once stable and mutable, shaped by time and folded into our consciousness. On the canvas, Cézanne unites both poles of the relation—the subjective and the objective—thus preparing the ground for the "revolution" of Cubism and, more broadly, for the transformations of *twentieth-century art*.

Art, for Merleau-Ponty, returns us to the things themselves. Painters, and artists more generally, have something essential to teach philosophers: they show us *how to see*. In Cézanne's work, Merleau-Ponty discovers a visual counterpart to what he was developing conceptually: a way of rediscovering perception and the order of meaning that emanates from things themselves. The nature Cézanne depicts is not *mimesis*; it is the process, the unfolding, and the genesis of form. His art is not imitation but recreation, an attempt to express the very becoming of creation, both visible and invisible. Gazes, movements, expressive forces, and rhythms inscribe themselves in the becoming of forms, recalling the life of the body and the world within the virtuality of Being. This is not a transcendent or pure Being, situated beyond phenomena, but a "wild," "raw" Being that intertwines, crosses, and transgresses, a Being that exceeds the visible and opens onto the invisible. The pictorial and the natural worlds thus belong to a single weave of relations and processes within reality, the very fabric that Merleau-Ponty would later explore in his ontological turn toward the *flesh of the world*.

Cézanne places before modern, and subsequently contemporary, art the problem of consciousness itself. By inscribing consciousness within the act of painting, he renders visible transformation, the becoming of the world through forms that express both disintegration and recomposition, the intertwining of invisible and visible forces acting within the phenomenal field and across the temporal arc of individuation. Forces understood as directions, vectors, and arrows find their pictorial expression in Klee's work as well as in the phenomenal field analyzed by the *Gestalt* psychologists.

Cézanne's mountain of Sainte-Victoire invites us, quite literally, to "*return* to the things themselves," that phrase which encapsulates the very spirit of phenomenology. The painted mountain is both the perspective upon the thing and the *thing itself*. The perspectival gaze that apprehends it is not a limitation but the very condition of appearance for a non-divine eye, an embodied gaze that cannot, by its nature, encompass the totality of the thing's unfolding at once. The limit, therefore, must be understood as an *opening* to the world. Things manifest only through their relation to a *perceptual system* (the body), which establishes the circular exchange of action and feedback, of world and meaning. In this sense, neither things nor world exist *in themselves*: both emerge through relation.

Through the analysis of the thing's appearing, the relation between phenomenology and the absolute knowledge of traditional metaphysics comes to mirror that between the thing and perception. The appearance of a partial aspect is not a mere adjustment to the thing; it must remain partial and limited in order to be total (cf. RC, 78). Cézanne's Sainte-Victoire reflects the *logos* of the sensible world that Merleau-Ponty sought to express: the circumscribed aspects of the visible that, precisely through their limitation, reveal the fullness of meaning contained in the thing's appearance. The painter's systematic return to the *thing* discloses that inexhaustible dimension of meaning which the work makes manifest. Both painting and thing belong to the same world. The painting becomes a generator of form and meaning, since the depicted object does not depend, as we have said, on *mimesis*; rather, it reveals an *equivalent* sense, a *virtuality* of reality's own appearing. The perceived world enables us to grasp the thing's virtualities as they take shape in a configuration at once stable and dynamic, a genesis of meaning expressed in the works of Klee and, before him, Cézanne. These are worlds emerging from a single process of reality that underlies the visible.

We tend to believe that we live among stable and autonomous things, yet in truth we inhabit a world of relations in perpetual becoming-structures that do not transcend the process, that are neither abstract nor atemporal, and certainly not arbitrary. The world we take as unique is but one among many worlds internal to reality. Such paradoxes arise from assumptions inherited from classical metaphysics and can only be overcome through a philosophy of relations, a relativistic perspective attuned to becoming. The world is the reverse side of human sensing: a gaze turned back upon reality that constitutes a world, even as this world grants meaning to our gestures. It holds within itself a sense that reality, at its most elemental level, does not possess; this sense arises instead from the reciprocal play between a form of life (the subject) and its environment (the world).

1.2 II

For Merleau-Ponty, phenomenology can be "practised and identified" as a *style* (PP, viii). The style of thinking is inseparable from the plane of phenomena it seeks to disclose and must serve as a guide. Through this style, which gives form to thought, the philosopher draws us nearer to the phenomena and the things themselves. To employ natural rather than formal language is to describe what we perceive, remaining anchored to the sense that phenomena hold within lived experience. It is precisely this plane of phenomenological appearance that the phenomenology of perception interrogates, analyzing the thing in accordance with the very conditions of its observability.

Engaging with Merleau-Ponty does not mean reading him in abstraction; it means turning our gaze toward what he himself paused to contemplate, those perceptual experiments in psychology that, now as then, retain their speculative power. Observing these sites in parallel with his texts brings us closer to the author than any purely conceptual study of his writings or life could do. Through his descriptive and speculative style, Merleau-Ponty communicates the most authentic sense of perception, teaching us, like a painter, to behold phenomena in their expressive plenitude. The task is to initiate a dialogue with him on the basis of a shared, intersubjective analysis of the visible and the invisible, together with a descriptive examination of appearance, in order to bring forth what has not yet been thought. This task becomes even more urgent when we turn to the fragments of the *Working Notes* collected posthumously in *The Visible and the Invisible*, a direction of inquiry that Merleau-Ponty had only begun to trace before his premature death at the age of fifty-three.

To think Merleau-Ponty anew, therefore, means learning to redirect our gaze toward the phenomena we wish to grasp, to look at the world according to its own rhythms, times, and modalities, without the impatience to subsume it under preconceived categories. It means receiving phenomena as they offer themselves, without subordinating them to the theoretical or practical demands of discourse. In short, it means letting the things themselves speak.

Immersing oneself in a philosophical work can captivate the reader, yet the intoxication of thought carries the risk of remaining enclosed within the text, bound to a web woven solely of concepts. The living observation of the world, however, exceeds the words that describe it. The breath that precedes immersion in the text must culminate in a new inspiration, born of fresh encounters with experience. For a thinker such as Merleau-Ponty—whose aim is to lead us beyond language and toward phenomena—this caution becomes a methodological principle. His texts must be read *in transparency*, allowing us to see through and beyond them, toward the world they strive to reveal. Pictorial art, too, can awaken our gaze, carrying it beyond signs and back to the taste of things themselves.

1.3 III

At the heart of Merleau-Ponty's reflections lies the body and perception, intended to awaken us from the "dogmatic slumber" that centuries of philosophical dualism have entrenched: the separation of mind and body, subject and object, thing-in-itself and phenomenon, true and false world, appearance and reality. All these oppositions dissolve once we rediscover them as woven into the same fabric of Being. Reading Merleau-Ponty means rediscovering the expressiveness of language as a reflection of the world itself: prose, like painting, becomes another *system of equivalence* with the order of meaning offered by the world. To move from language toward the spectacle of the world is not a one-way passage; it is a way of letting phenomena speak. The world should not be conceived as a vanishing point external to language, but as a *frame of reference* implicit in our bodily and linguistic behavior: the *background* that lends meaning to the figures of our practical and verbal engagement with things.

Rediscovering the phenomena described by Merleau-Ponty is thus a fertile preliminary task, one that allows us to enter into dialogue with him, while taking as our reference the phenomenal present as it takes shape here and now. We perceive the world around us in its multiplicity of forms, colors, scents, lights, and shadows, all of which compose the "spectacle" that stands before our eyes and envelops us, even as it continues to exist behind our backs.

Perception is not a substance perceiving itself, nor are we ordinarily conscious of consciousness as such. What is directly perceived are phenomena that manifest both outside and within the body. At times they appear endowed with greater concreteness and stability; at others, as in the case of "afterimages," they reveal their ambiguity or a certain "subjective" quality. Yet in no instance do we step outside the field of immediate experience, whose description requires no reference to the underlying neurophysiological processes. Leibniz had already observed in the *Monadology* that direct perception is distinct from the transphenomenal mechanisms that sustain it. The physiology of the brain may be likened to the gears of a machine: it is necessary for its functioning, but never sufficient to explain the immediacy of lived experience, which remains the proper domain of the phenomenology of perception. This does not reintroduce a dualism but rather distinguishes perception as lived in the first person from its subsequent representations. Nor is phenomenology concerned with formulating a definition of perception; rather, it teaches us once again how to observe the world, to educate our gaze to apprehend the meaning that things hold for us.

The relation between perception and science corresponds to that between appearance and reality (cf. *C*, 14). These equivalences—perception/appearance, science/reality—emerge with Descartes, who first posited the *cogito* as the most indubitable source of knowledge. From that moment onward, the "world" became a possible "object" of doubt, and the correspondence between the contents of the *ego cogito* ("objective being") and external reality ("formal being") could no longer be presupposed but had to be demonstrated, a field of theoretical conquest rather than immediate evidence. With the advent of the *cogito,* the West acquired a new awareness of the possibility of doubting the external world. This inaugurated a relation of dependence between external phenomena and the mind, giving rise to a renewed problem of *adaequatio* between the *cogito* and the thing itself. Yet within the life of consciousness we immediately experience both the activity of thought and the perception of things and events; both belong to the same order of phenomena, whether lived as internal or external. The meaning of perception may appear self-evident, since it has always mediated our access to the world. As Kant reminds us, all knowledge begins with experience. Yet it has taken centuries of philosophical and scientific inquiry, from the metaphysics of antiquity to contemporary neuroscience, to clarify its nature, and even so, a residue of opacity endures.

Although Merleau-Ponty engaged with the entire history of philosophy, his most rigorous historical–conceptual dialogue begins with Descartes, whose thought he subjects to a critical analysis in order to uncover the origins of modern philosophical and scientific thinking. This constitutes a decisive moment whose essential features still reverberate in contemporary conceptions of subjectivity, world, and reality. The Greek notion of *substratum* was reformulated in modernity as *subjectum,* as the foundation of knowledge and the site from which the world is apprehended. With Descartes, the implicit realism of the Greeks gives way to the idealist turn of modern thought. Yet Merleau-Ponty traces a different path, sketching an alternative understanding of phenomenology that transcends the polarity between realism and idealism. The domain of phenomena is conceived not through the opposition of subject and object, but through a relational perspective.

According to Descartes, to grasp the unreliability of the senses it suffices to observe how things appear, thereby turning to the intellect as the ultimate ground of indubitable knowledge (*episteme*). His well-known example is that of the piece of wax: perception cannot apprehend its essence through observation alone. We perceive the unity and identity of the object, yet when heated these properties vanish. Faced with the melted wax, we can affirm its sameness only through the intellect: the qualities have changed, while the substance remains. For Descartes, perception marks the beginning of a science that is still *confused*; yet this very *vagueness* conceals an epistemic advantage, a heuristic power that Leibniz would later describe as the domain of "clear and confused" ideas, in contrast to the "clear and distinct" ideas that define conceptual knowledge. We might, for instance, glance at a table and immediately sense that there are *several* objects upon it without knowing their exact number: such "good vagueness" suffices for our everyday comportment and ensures coherence in most of our actions, which rarely demand analytic precision. The body, understood as a perceptual system, thus provides the first measure of orientation within the surrounding world.

1.4 IV

Phenomenology, in all its articulations, maintains an irreducible space of reflection between scientific knowledge and our subjective life, a remainder that cannot be absorbed by scientific explanation. It is within this interval that Merleau-Ponty situates his thought. In this sense, he never sought to repudiate science; on the contrary, his work unfolds in continuous phenomenological dialogue with it.

Let us consider the relation between appearance and reality. With the advent of modernity, experimental science assumed the task of investigating the reality that underlies experience. The problem does not lie in this aim, but in the way we interpret the relation between perception and appearance. If, in the pair appearance–reality, we place "reality" on one side, then on the other we are left with something merely apparent. The world, insofar as it depends on perception, that is, on those qualities that Galileo called "secondary" or "subjective," would thus cease to belong to reality, since such qualities vanish with the absence of the subject. Yet, as the empiricists (above all Berkeley and Hume) observed, the so-called primary qualities are likewise objects of experience. If we may doubt secondary qualities, we may equally doubt the primary ones.

The meaning of the relation between appearance and reality must therefore be critically reexamined through a phenomenological analysis of the reference systems involved. Otherwise, we risk construing perception as mere illusion opposed to reality. If we maintain the pair science–reality (with reality as the object of scientific investigation), must we also retain the pair perception–appearance? If science can study reality, can the study of perception not also constitute a legitimate scientific inquiry? To overcome this impasse, we must clarify the sense of "reality," "appearance," and "perception," and elucidate the kind of relation science entertains with philosophy. It is precisely on this set of questions that Merleau-Ponty centers his early reflections, questions that are not preliminary but essential to the trajectory of his thought. What seem introductory themes lead, in fact, to the very core of his phenomenology.

Since its object is the *Lebenswelt*, phenomenology cannot restrict itself to a mere description of appearances; it must integrate this descriptive level with a scientific and experimental analysis. At the same time, phenomenology is also the method that, after the triumphs of modern science, exposes the "crisis" of scientific knowledge, showing that the experimental method has led us toward a model of "objective" knowledge that neglects the subjective and qualitative dimensions proper to the life-world and its multiple meanings. Science takes the world as appearance on the assumption of an independent reality, whereas philosophy has long shown that knowledge begins from experience and that the very assumption of external reality is far from self-evident. Science presupposes reality in order to conduct its investigations; philosophy, by contrast, presupposes nothing uncritically. It must interrogate "world" and "reality" as the very presuppositions of science. To investigate them, philosophy must first *suspend* judgment about their existence, for not all of reality can be given in experience, only that portion we call "the world." The distinction between world (the correlate of experience) and reality allows us to problematize the relation between science and philosophy.

Merleau-Ponty devoted sustained attention to the sciences, and above all to psychology, especially Gestalt psychology, understood as a "science of observables." World, reality, perception, art, and knowledge, both scientific and phenomenological, constitute the principal threads of his reflection. The appearing of the world, far from containing "contradiction," is perfectly consistent with the principle of non-contradiction, which governs the manifestation of things and the world. Even in the case of bistable figures—such as the famous face–vase image of Gestalt psychology—we encounter a constraint within appearing: one figure excludes the other, each manifesting at different times. "Contradictions," like truth, concern the order of judgments and classification of the observed phenomena, not the level of appearance itself.

Appearing does not share the same conceptual nature as thought, since it cannot be falsified, even when it manifests as hallucinatory, subjective, confused, or ambiguous. What is required is to discern the reference systems underlying the "logic of the sensible": phenomena in their very modes of manifestation, without reducing them to language. In the phenomenology of perception, the conceptual level cannot claim priority over appearing; rather, it is the modes of appearing that establish the "rules of the descriptive game." In relation to language, the study of perception enables us to escape dichotomous logics. Merleau-Ponty leads us toward a form of thought made of nuances, where one no longer thinks in "black and white" but attends to the multiple gradations of the visible world. Wolfgang Metzger offered a definition of phenomenology that Merleau-Ponty would have appreciated: to practice phenomenology, one must accept that what is given may appear illogical, senseless, and contrary to our expectations or habitual modes of thought. We must "let the things themselves speak," without allowing the limits of vocabulary or logic to sever us from them.[2]

To investigate the sciences phenomenologically is thus to relate them to the meaning they hold for us. This is not to reduce physical reality to the world, but to clarify reality (ontology) through its internal relations, *one* of which is the subject-world relation (epistemology). The meaning the world unfolds not only orients our actions but transforms their significance: folding back upon us, it reshapes our behavior and our relation to things. The world, to be such, requires perception (subjective pole), it requires a perceiving body. In other words, the "world" emerges as the term of a *complex*, anti-reductionist relation: the expression of the subject–world structure (objective pole).

[2] See W. Metzger, *Psychologie*, Darmstadt: Steinkopf 1941.

Thought, however abstract, arises from the perception of the world, from that sensible dimension which language names but does not create, from that place where language itself takes form. Art, too, is born from this same relation that shapes our *being-in-the-world*. We become aware of things at the moment we gain consciousness of them; our existence begins there, though something has already seen and thought *before* consciousness intervenes. Our body participates in the *entire* process of perception. All this occurs "out there," yet this "there" exists only in relation to a "here" given through our body, a hinge, a living connection between consciousness and world. Our being in reality binds us to the process of emergence, to that which arises from the "world" through the interaction between the *Leib* (the lived body) and the *Körper* (the physical body): "Our own body is in the world as the heart is in the organism: it keeps the visible spectacle constantly alive, it breathes life into it and sustains it inwardly, and with it forms a system" (PP, 235).

In Klee's paintings we can visually apprehend the very sense of generation. The forms of existence of the "thing" are no longer considered in their stability, as finite, immobile configurations, but are referred back to their *process of formation*: the dynamic that encompasses both the genesis of artistic creation and the meaning of the world for us. What is the relation between the world as lived and the world revealed in painting? And between this world and the *virtual* worlds we inhabit today? How can we give voice to sensible experience without falling back into the dichotomies of traditional metaphysics? These aporias belong to the *logos*, which crystallizes truth in the attempt to fix becoming through immobile forms of thought. It is precisely the critical analysis of the concept of *form* that constitutes the legacy Merleau-Ponty bequeaths to Gilbert Simondon, beginning with his reflections on "nature," a legacy that Simondon develops in his own theory of individuation, inspired by Merleau-Ponty's final lectures at the Collège de France.[3]

The focal point of our inquiry lies in showing that the perceived world is, in non-arbitrary terms, both "origin" and "meaning giving", an implicit frame of reference for every subsequent articulation of knowledge, whether scientific (from world to reality) or plural (a multiplicity of worlds cohering within a "system of equivalences" grounded in *this* world). We can grasp such movements of thought by crossing "verticality" (as the element of transcendence) with the *horizontal plane* of experience (the plane of immanence): the world as horizon of meaning and as reflection of a reality that is not directly visible. The task is to thematize the "point of beginning" of knowledge, starting from what we observe, from action, from others, from the inner and outer world, in search of a synthesis capable of expressing the very sense of our intertwining with this dimension of Being.

[3] See especially G. Simondon, *Cours sur la Perception (1964–1965)*, Chatou: Éditions de La Transparence, 2006.

1.5 V

The description and understanding of phenomena entail an analysis of the implicit frame of reference shaped by our corporeality (high and low, small and large, near and far), an analysis that functions according to the scale of the phenomena under consideration. Phenomenology outlines a mode of thought devoted to questioning things in their tacit presence, a form of thinking capable of grasping the originary source of meaning inscribed within perception. We find this descriptive level in Merleau-Ponty's analysis of behavior in *The Structure of Behavior*, a work that became an essential point of reference for both continental and Anglo-American debates. His early writings seek to "rediscover the world of perception" and its theoretical implications, while his later works aim to show how critical thought surpasses the perceptual dimension that has "initiated us to the truth" (PrP, 3). From *The Structure of Behavior* emerge many of the themes later developed in his major 1945 work, *Phenomenology of Perception*, and deepened further in his final, unfinished project, *The Visible and the Invisible*. These three moments delineate three phases of Merleau-Ponty's thought. In *The Structure of Behavior*, the relation between consciousness and nature, understood in its organic, psychological, and social senses, is analyzed from an external standpoint, through the detached gaze of the *spectateur étranger*. In the *Phenomenology of Perception*, the same relation is examined from within, articulated and problematized through an analysis of the life of consciousness. The final phase represents the attempt to elaborate a new ontology, one in which even the element of consciousness proper to phenomenology is itself brought into question. Each of these moments articulates a distinct way of intertwining perception, world, and reality. A separate chapter could be devoted to Merleau-Ponty's political reflections, developed in works such as *Humanism and Terror* (1947) and *Adventures of the Dialectic* (1955), as well as in his contributions to *Les Temps modernes* until his resignation in 1952, when he publicly broke with Sartre.

Perception remains the axis around which Merleau-Ponty organizes his attempt to renew philosophy. Having studied both Gestalt psychology and Husserlian phenomenology, drawing from them fundamental notions such as perception, experience, and form, he undertakes a critical rethinking that fuses and transforms both traditions. His work is not a rupture with Husserl but an expansion: Merleau-Ponty explores Husserl through Gestalt psychology and rereads Gestalt thinking through conceptual tools derived from Husserl. His tireless excavation of Husserl's texts is enriched by dialogue with fields beyond philosophy: literature (Proust), painting (Cézanne), linguistics (Saussure), anthropology (Lévi-Strauss), and the natural sciences, particularly physics and biology.

Merleau-Ponty knew that his project stood in opposition to a long metaphysical tradition that grounded knowledge not in perception but in the *cogito*. With the *cogito*, modern subjectivity is established, and with it, the Cartesian dichotomy between subject and object, a divide Merleau-Ponty seeks to dismantle by restoring to the body and perception their intrinsic ambiguity. This dismantling unfolds through a shift from *cogito* to perception: "the thinking subject must have its basis in the subject incarnate" (PP, 225). Merleau-Ponty thus affirms the primacy of perception, taking the visible appearing of the thing as prior to linguistic and conceptual categories. To thematize something prior to such categories, while acknowledging that one cannot dispense with language, is a theoretically delicate operation. Things appear "as they appear," manifesting themselves tacitly and opening onto a horizon of meaning: they are "already made by me before any analysis" (PP, xviii). Silence and speech, figure and ground, the visible and the invisible—these are complementary aspects that, throughout Merleau-Ponty's work, imply and sustain one another.

The task, therefore, is not to justify but to understand and embrace the primordial dimension of meaning that constitutes the world's appearing, of which science offers only a "second" image, one never free of metaphysical presuppositions. Merleau-Ponty seeks to construct a phenomenology that addresses not only the essence of things but their existence. To ask how the aspect of a thing relates to the *thing itself* and to its conceptual definition is to hold together the essence and existence of the thing. The perception of a given object does not presuppose its concept as a condition of visibility. The concept of "thing" may refer either to its intrinsic properties or to the class of its appearances, and both senses acquire meaning only when anchored to a determinate perspective. We have experience of things prior to any conceptual definition, and this experience grounds the universality that phenomenology attributes to the *thing itself*. Phenomenology thus aspires to a universal form of knowledge rooted in the radical problematization of place: it implies a concrete, situated stance within the world. Perception is an original experience, the condition for the thinkability of the thing itself. It is on this transcendental dimension that *The Structure of Behavior* concludes, and from which Merleau-Ponty's subsequent philosophy takes form.

Merleau-Ponty credited the Gestalt school with having placed the notion of *structure* before that of meaning, thereby revealing "the contingent arrangement by which materials begin to have meaning in our presence, intelligibility in the nascent state" (SB, 206–207). Beginning from the analysis of perceptual behavior, he directs his critique against the Cartesian paradigm that separates thought from corporeality and more broadly, against the dominant epistemological paradigms of his time. He challenges their overreliance on causal explanation and the objectifying tendencies of reflexology and behaviorism, as exemplified in the works of Pavlov and Skinner. For Merleau-Ponty, the most adequate theory for explaining perception and behavior is found in *Gestaltpsychologie*, whose concept of "form" offers a different epistemic model: physics, he argues, derives its sense only in relation to the *form* of things, a form that is not a physical entity but an object of perception (cf. SB, 135). The perceptual dimension is thus prior to intellectual categorization and to the distinction between the "in-itself" and the "for-itself." The notion of form marks a decisive turning point: it is neither idea nor thing but a functional structure that allows the human being to organize behavior in the world, preserving the structural ambiguity that binds consciousness to objective exteriority. A further stage of Merleau-Ponty's project is to move beyond the *psychological conception of Gestalt* toward a *philosophy of Gestalt*, in order to avoid the trap of psychologism, the view that knowledge is merely a part of being and an imitation of things (cf. SB, 135).

To paraphrase Heidegger, the works of great philosophers represent nothing other than the full unfolding of a single fundamental idea around which all else revolves. In this case, our concern has been to highlight the essential traits of the "phenomenology of perception" as the gravitational center of both the phase preceding and the phase following the publication of the book that bears its name. Such a focus, we hope, can illuminate the coherence of Merleau-Ponty's entire philosophical trajectory.

1.6 VI

Just as the first lines traced on a sheet determine the subsequent unfolding of a drawing, so too the premises of our philosophical discourse condition the movements of thought that follow: the mode and direction of our questioning guide and delimit the possible response. The beginning of knowledge resounds with this same necessity and must itself be problematized. Phenomenology grasps and embraces the world's appearing in its dynamic and expressive dimension. Unlike empiricism, it does not aim to isolate the ultimate elements of the thing nor to reduce it to representation; what manifests itself must be received as it gives itself, in its fullness and *complexity*. The beginning, whether perceptual or verbal, does not arise from nothing, nor is it the capture of a given instant, a photographic snapshot of being. We always perceive something that *points beyond* itself. In the *phenomenal present* of ongoing observation, aspects of things announce themselves even without a corresponding physical stimulus: we live in a continuum of relations and processes, not in a world composed of static frames. First-person knowledge takes form in the moment we grasp a situation *hic et nunc*, yet this knowledge is never absolute: it is shaped by language, culture, and the sedimented history of our existence. The initial situation must therefore become the object of phenomenological inquiry, a processual analysis of decomposition and recomposition, moving from the immediate to the mediated, from the whole to the parts and back again. The beginning is not an instantaneous point without extension, nor a polarity of being and nothingness; it is a *gesture*, a complex event endowed with duration, a "thickness" of existence.

If there is a grammar proper to perception, it may serve equally to clarify the forms of thought. In this sense, phenomenology is also a work of conceptual clarification, addressing the tacit presuppositions of knowledge: the silent manifestation of the world and of things that precedes and receives our words and concepts. Even the simplest act of perception, a glance at a small portion of the world, implies our belonging to the *Lebenswelt*. To grasp this "phenomenological heritage" is to return to the origin, to that which generates and stabilizes the meaning of the world and its forms. Such reflection leads inevitably to the question of the beginning of knowledge, and particularly of scientific knowledge. For Merleau-Ponty, this entails a "return to the phenomena," a "return to the things themselves," and to that foundational gesture which Husserl called *epoché*, the suspension of judgment concerning the existence of the world. Though reinterpreted, this gesture continues to play a decisive role in Merleau-Ponty's own phenomenology. His reflections seek to understand the sense of *this* world, which, as Nietzsche might say, can be fully justified "only as an aesthetic phenomenon," that is, as perceptual. Humans bear within themselves, silently, the paradoxes of philosophy because metaphysics takes as its object the experience of the world, of others, of history and culture, those elements that secretly operate before any concept or act of consciousness (cf. SNS, 94). This "beginning" is indissolubly bound to perception, understood as a field of meaning and existence. The event of "secrecy" designates this preconceptual dimension that, beneath the surface, acts upon our body and our relation to the environment. There operates within us a form of nonconceptual, visual thought, a "thought of the body," the product of millennia of evolution, a continuous exchange of information between body and world (cf. N, 222). A "perceptual faith" orients us toward a primordial order of meaning. The body is an integrated, incarnate system within the environment: "Philosophy" writes Merleau-Ponty, "is the perceptual faith questioning itself about itself" (VI, 103). In his final phase, he names this relation "faith" because, like every faith, it implies doubt and continuous questioning. Yet this "faith" concerns the world as given in its appearing, as something that seems to exist independently of us. It is precisely this faith, characteristic of naïve realism, of the natural attitude, and of scientific objectivism, that must be examined and brought to reflection. Merleau-Ponty speaks of "perceptual faith" to designate those elements that lead us to take the certainty of the world as self-evident, grounded on the features of appearing unveiled by *naïve physics*.

To practice philosophy means to question, at least once in one's life, as both Descartes and Husserl insisted, all of one's certainties. Ideas cannot be truly grasped without transforming our fundamental convictions; otherwise, learning remains external, adding notions without altering vision. To adopt a new perspective is not to accumulate ideas but to challenge those incompatible with it: it is to work upon oneself, upon one's way of inhabiting the world. In our uncritical commerce with things, in the everydayness of being-in-the-world, we are "naively" thought by the very language we use: we inherit the history and culture that precede us (cf. S, 90). To grasp the meaning of the world requires *distance* from the conceptual filters of language. A specific historical process has shaped the grammar through which we think and name the spectacle of the visible. In this sense, we may say paradoxically that we are spoken by language before we speak through it. Language cannot be regarded as a (neutral) medium between us and the world: it filters and shapes experience. Hence Merleau-Ponty's difficulty in recovering a prelinguistic dimension, already thematized by Husserl in his notion of *Lebenswelt*, which Merleau-Ponty reinterprets and expands within his own phenomenological project.

The word, like the *gesture*, is not an isolated element. Even the simple act of pointing cannot be understood as the primitive building block of language (cf. S, 67). If it is to be grasped in its full expressiveness, the gesture must be seen not as part of the body but as an act of the whole body within a complex system, just as a word finds its meaning only within the sentence. The meaning of a gesture cannot be reduced to a physical or physiological function but must be understood phenomenologically, in light of the network of relations in which it is inscribed (cf. PP, 216). The expressiveness of the word shares the same irreducibility as the gesture of the hand: both presuppose our acceptance of the world and of the body as implicit frames of reference.

We thus return to the theme of the appearing of the thing, reintroduced through successive stratifications: just as repeated brushstrokes shape the figurative space of the canvas, we revisit each aspect of discourse in a continual renewal, moving toward that single, ever-unfolding discourse which, like a Möbius strip, never closes upon itself but perpetually finds its *beginning anew*. Like threads intertwining in a weave, phenomena compose the spectacle of the world and the structures of meaning that host our behaviors. Such phenomenological observations do not crystallize into fixed structures but unfold within a dynamic field that continuously seeks unity and *stability of meaning* amid constant transformation: "We shall find in ourselves, and nowhere else, the unity and true meaning of phenomenology" (PP, viii).

To read Merleau-Ponty through the lens of the Gestalt tradition (and through the notion of immediate experience) is thus to take a path with "heretical" traits in relation to Husserlian orthodoxy. The task is to problematize the very beginning of knowledge, situating these reflections alongside the questions that preoccupied Merleau-Ponty himself: the relation between meaning and perception, whose clarification can illuminate our existence, the world, and its contemporary transformations. At the heart of our inquiry lies the conviction, shared with Merleau-Ponty, that *meaning* is not the mere product of consciousness but the result of the intertwining of subject and world.

Chapter 2
World, Environment, and the Meaning of Dwelling

"Science manipulates things and gives up living in them" (EM, 159). With these words Merleau-Ponty opens his final work, *Eye and Mind*, a dense and decisive text in which the theoretical trajectory of his late philosophy becomes fully visible. Scientific thought, oriented toward the comprehension and manipulation of reality, neglects the distinctive features of *subjectivity* and, with them, the very *sense* of our dwelling in the world. What is the relation between the manipulability of things (the transformations effected by the techno-sciences) and the meaning we ascribe to the world and to our ways of inhabiting it?

The manipulative possibilities emerging from technological applications of science presuppose abstract operations and the mobilisation of symbolic resources. The use and transformation of things arise, first of all, from that part of our body (the hands) freed to act upon the surrounding environment. By attaining an upright posture, the human being also secured the evolutionary advantage of possessing limbs released from locomotion and capable of reshaping the world. The meaning of the environment (or world) can thus be conceived as that which surrounds every living being: a portion of experience that may, in part, be "shared" across species. Relative to the human world, certain regions of experience overlap with those of other forms of life; yet each living being inhabits its own world. We cannot have direct access to another's world, human or non-human. The logic of the other's behaviour is inferred from what appears coherent within our own world. Such coherence presupposes a single reality underlying a plurality of possible worlds.

Different forms of life entail, as Jakob von Uexküll showed through the notion of *Umwelt*, later taken up by Merleau-Ponty (cf. N, 171), the possession of a "surrounding world" or "environment-world." The interplay of shared invariants and interspecific differences gives shape to this concept: the identification of worldly constants, constraints, and conditions of emergence, together with the analysis of behaviour across species, carries both epistemological and ontological significance. On one hand, we must affirm the non-coincidence between the *physical world* and the *environment-world*, the latter being apprehended immediately and describable

L. Taddio, *The Appearing of Sense*, Lecture Notes in Morphogenesis,
https://doi.org/10.1007/978-3-032-27021-4_2

phenomenologically. On the other, we must relate, and yet distinguish, the notions of *world* and *reality*. Animals endowed with different "image-apparatuses," a term from Konrad Lorenz, the father of ethology, possess distinct worlds while sharing a single reality (monism). Each "world" arises from specific relational processes; for Merleau-Ponty, it constitutes a structure that must be grasped as a body–subject–world system. Since the world is the correlate of experience, there can be no world without a subject, just as there can be no subjectivity without a world.

Starting from the notion of *affordance*, science (*episteme*) expands our capacity to manipulate and transform the world, thereby altering the meanings we ascribe to things. The surrounding environment is never neutral; it is saturated with meaning and orients our conduct within the field of perception. The concept of *affordance*, which partially continues and partially replaces the Gestalt notion of "field," was developed by James J. Gibson, founder of the ecological theory of perception, drawing on the work of Kurt Koffka. It reveals how things, through their intrinsic expressive properties, present themselves as potential invitations to action. Having grasped the expressiveness of the world and its philosophical import, Merleau-Ponty can be regarded as one of Gibson's precursors.

Contemporary science has reached such a degree of abstraction that the scientific image of reality no longer bears resemblance to the world of experience. This enables an ever more radical transformation of the environment, culminating in the creation of digital worlds that are infinitely malleable. From what we are able to do, the practices of everyday life derive their use-value: the manipulable aspects of things define the technical-practical horizon of our action, appearing as means toward specific ends. Yet while phenomenology can analyse how we use things, Merleau-Ponty directs our attention first to recovering the sense of things as they appear to us. To rediscover their aesthetic–perceptual dimension means returning to a precategorical level that precedes functional use. For Merleau-Ponty, the rediscovery of metaphysics thus entails giving voice to this silent dimension while simultaneously enacting a partial *withdrawal* from the natural attitude of everyday life. It means interrupting the uncritical praxis that sustains our habitual engagement with the world and re-educating our gaze to perceive how phenomena manifest themselves, to recover that sense of *wonder* from which philosophy itself arises. Phenomenology, in this sense, involves both a suspension of practical absorption in daily life and a *suspension* of the manipulative attitude toward the world, an attitude shaped by the imperatives of scientific and technological progress.

However, much science may expand our capacity for intervention and manipulation, the order of meaning remains rooted in the qualitative dimension of experience. When meaning itself becomes an object of transformation, it opens to an infinite potential for manipulability. Following Husserl and Heidegger, Merleau-Ponty interprets technological development as a quantitative representation of reality: even the digital world, constructed through discrete logic, is nonetheless lived through interfaces that render it accessible to our experience. To say that "science renounces dwelling" is to recognise that lived space does not coincide with the space of geometry or physics. The meaning of dwelling can only be understood phenomenologically, that is, through the constitutive role of the body and subjectivity. Thus, for example, what we call a table, in the physical sense, possesses none of the qualities that make it, in our experience, *this* table we use in daily life.

We perceive the table according to the characteristics that make it meaningful for us: as a surface for support, of medium size, suited to our corporeal scale. Means and ends are bound to our form of life. A thing may be "large," "small," "distant," "opaque," or "smooth" only insofar as it is apprehended relative to our embodied condition within a concrete situation. This implicit frame of reference determines how we evaluate and act within the (lived) surrounding environment. In the absence of gravity, for instance, "up" and "down" lose their meaning: our body, in a non-arbitrary manner, establishes within the phenomenological field the relations that depend on the situation in which it is immersed.

Chapter 3
On Exactitude in Science

Let us consider the example proposed by the astrophysicist Arthur Eddington. On the one hand, there is the table as it appears to us in phenomenal experience, an object familiar and intimate, integrated into the fabric of our everyday life. On the other, there is the physical table, understood as that which exists in itself beyond appearance: a configuration of atoms, electrons, and particles described by the theoretical framework of quantum mechanics. In the scientific image of the thing, Eddington observes, we no longer find the table as such, in its unity and identity. There is no correspondence between the stable, colored, and tangible table we perceive and the swarm of moving particles that physics reveals as its "true" composition.

Our scientific knowledge of the nature of the table has thus distanced us from the "something" we see and experience, the object as lived and represented. Science and experience appear irreconcilable, like paths that diverge indefinitely. Yet reality is one, and the world is part of reality: phenomenology must therefore assume the task of reestablishing the conjunction between reality, world, and the lived practices of life. This demands an analysis of multiple frames of reference and the virtualities of appearing, grounded in the phenomenological distinction between appearance and reality.

A similar shifting of planes occurs in the notion of *image*: the image "stands for" something, in that it refers to or replaces it. The scientific image of the world "stands for" reality, while the world itself is an appearance structured by our image-apparatus. For a long time, the world was synonymous with reality; now it risks reduction to a mere image, a subjective construction dependent on how the brain processes information. Yet the meaning of existence remains rooted in the sense we ascribe to our environment and to the places in which we dwell. Transphenomenal reality, being inaccessible to direct experience, is instead expressed by science through mathematical symbols and physical laws that describe a reality either too vast or too minute to be perceived.

In *Del rigor en la ciencia* (*On Exactitude in Science*), Jorge Luis Borges recounts the story of a people devoted to cartography who perfected their art to the point of

L. Taddio, *The Appearing of Sense*, Lecture Notes in Morphogenesis,
https://doi.org/10.1007/978-3-032-27021-4_3

Fig. 3.1 Mach's square

producing a map of the Empire at a scale of 1:1, corresponding point by point to the actual world. Later generations abandoned this art, Borges writes, recognising that such maps were useless; only fragments of them remained. With masterful concision, Borges shows that an image, to function as an image, must differ from what it represents: were it to coincide completely with the thing, it would cease to be an image. The image "stands for" the thing by differing from it. A similar principle applies to consciousness, conceived as the mental image of the world: it must differ from reality in order to be a world—an image coordinated with reality. To return "to the things themselves" thus means to return from *reality*, as grasped by science, to the *lifeworld*, the place in which we dwell and project our existence, the origin of meaning and of all aesthetic valuation.

To clarify the meaning of *dwelling* and the phenomenal perception of things, let us consider the so-called "Mach's square." The second figure, from a geometrical standpoint, is identical to the first, merely rotated by 45°. In geometrical space (science), the two figures are perfectly congruent; yet phenomenologically (in experience), they appear different. While geometrically the same, the rotated figure assumes another appearance and meaning: most observers perceive it as a diamond or rhombus (Fig. 3.1).

From the standpoint of the intellect (Merleau-Ponty's *intellectualism*), we affirm the figure's identity: "a square is always a square, whether it stands on a side or an angle" (PP, 54). From a phenomenological perspective, however, the two figures are "barely recognisable" as the same: science and experience are not superimposable; their orders of meaning do not coincide. Vertical and horizontal axes play a decisive role as coordinates within the environment in which we move, beginning from the phenomenal body–world frame of reference. In Mach's square, the angles of the second figure coincide with these axes, forming vectors that "point" in "meaningful" directions, for they constitute the reference coordinates of our embodied action. Similar considerations are found in the experiments of von Schiller and Wertheimer.

In geometrical space, by contrast, there are no privileged directions, no angles endowed with expressive significance. The lived environment, however, possesses qualitative, expressive features that no map, as a scientific representation of space, can convey, since its function is of a different order. For Merleau-Ponty, scientific knowledge stands to things as *geography* stands to *landscape*: an abstract relation in which the former depends upon the latter. A map relates to the environment differently than perception relates to the landscape; yet to fulfil its function and possess meaning, it must implicitly refer back to the body that reads and interprets it.

The environment that surrounds us is unmediated: perception inaugurates our opening to the world. "To return to the things themselves"—the motto of phenomenology—"is to return to that world which precedes knowledge, of which knowledge always *speaks*" (PP, ix–x). Merleau-Ponty's comparison between geography and landscape recalls a well-known story retold by Kurt Koffka: one winter evening, a knight arrives at an inn after a long journey. When asked from where he has come, he points to the plain of snow he has just crossed. To his astonishment, he is told that he has traversed the frozen Lake Constance; struck by this revelation, he falls dead on the spot.

Koffka recounts this German legend in his *Principles of Gestalt Psychology* to distinguish the *physical environment* (reality) from the *behavioural environment* (the phenomenologically lived world). Learning and basic adaptive behaviours occur only within this latter level, which we designate as *world* to differentiate it from *reality* understood as what is not directly observable, the object of scientific investigation. This distinction does not imply dualism, for the diverse environment-worlds proper to different forms of life correspond coherently to the same underlying reality: they are emergent properties of the relations among living bodies (*Körper*).

If we conceive the world as a complex system structurally bound to our subjectivity, and conversely understand subjectivity in Gestalt terms (as structure) in relation to the world, then, from the standpoint of classical realism, the question of whether a world exists independently of us becomes meaningless. This clarifies Merleau-Ponty's insistence on our "being incarnate in the world." To ask whether the external world could exist apart from subjectivity is as nonsensical as asking whether water, given the formula H_2O, could exist without hydrogen or oxygen: it is a contradiction in terms. Some emergent properties we attribute to water, such as transparency or clarity, may also belong to other substances, but to classify phenomena empirically is not the same as understanding the conditions of their appearance.

The relational dynamics that allow a phenomenon to emerge, and the identification of its dependent and independent variables, require experimentation to be recognised; they cannot be grasped *a priori* through an eidetic or merely descriptive method. This in no way diminishes the value of phenomenological description, nor the validity of other methods. One may adopt a "relativist" perspective without excluding the question of "truth;" on the contrary, to apprehend reality we must first define our investigative tools and render them adequate to the level of reality under examination.

Referring to Einstein, Merleau-Ponty notes that within the perceptual domain there are phenomena that, taking our body as a frame of reference, may be described as simultaneous (cf. SNS, 55). The same applies to our perception of the Sun's apparent motion relative to the Earth: since we do not directly perceive the Sun's displacement, our judgment rests on experiential inference, an observation corroborated by studies in so-called *naïve physics*.

In his *Autobiographical Notes*, Einstein admits that he may appear to systematic epistemologists as a kind of unscrupulous opportunist, since, depending on the situation, he can assume the attitude of a realist, in investigating a reality independent of experience; of an idealist, in treating theories and concepts as free inventions of the human mind; of a positivist, in grounding such inventions on logical relations among experiences; and of a Platonist, in upholding logical simplicity as an indispensable criterion of research.[1]

From this perspective we draw the idea that reality cannot be apprehended from a single principle, methodology, or epistemological premise. It is the nature of the problem that determines the appropriate method: one does not begin with the method, but with the problem itself. The dream of the old metaphysics, to account for the whole from a single standpoint, has shown its limits. Depending on our investigative aims and on the plane of reality in question, we must adopt distinct methods and assumptions suited to each. The phenomenological method, while inadequate for grasping *reality* in itself (for reality cannot be deduced directly from the observable world), is eminently suited to the apprehension of the *world*.

From an evolutionary perspective, we may understand that the primary function of the *image-apparatus* is not to uncover the laws of the universe but to respond to adaptive needs: if the world were incoherent with respect to the actions of a given form of life—a term we employ here although it lies outside Merleau-Ponty's lexicon—that form of life could never have adapted to its environment. Gestalt theorists such as Wolfgang Köhler identified a fundamental *isomorphism* between neural, perceptual, and physical structures, a problematic notion, yet one that continues to provoke questions and open perspectives still relevant today.

From this evolutionary standpoint, even the contents of thought (concepts themselves) participate in our apprehension of the world, for thought arises from experience and constitutes its articulation. As Merleau-Ponty affirms, we are made of the same *étoffe*, the same fabric of Being. How, then, could thought, could the *cogito* emerge apart from the body and from experience? Is not the absence of an embodied orientation toward the world, an incarnate relation to Being, perhaps the greatest limitation of our present attempts to create a genuine form of Artificial Intelligence (AI)?

[1] Cf. A. Einstein, *Einstein on Einstein: Autobiographical and Scientific Reflections*, Princeton: Princeton University Press, 2020, p. 152.

Chapter 4
Phenomenology and Science

Although phenomenology plays a crucial role in offering internal clarification within the sciences, it is not self-sufficient for the investigation of reality. If we examine Merleau-Ponty's working method, as it emerges, for instance, from his 1953 course at the Collège de France, *The Sensible World and the World of Expression*, we find a mode of inquiry that, in practice even before theory, confirms this point. Merleau-Ponty demonstrates his capacity to work within experimental psychology, that is, within the scientific domain itself, in order to draw out its epistemological and ontological implications.

Phenomenology, by recovering the order of phenomena, articulates the *grammar of the world*, not of reality. Merleau-Ponty's preference for descriptive forms closer to literature than to strictly logical-formal discourse accords with the expressive character of phenomena and with the modes through which they offer themselves to consciousness. The dimension of meaning in appearance, grasped in its entirety, finds its most adequate expression in natural language, capable of conveying the nuances inherent in the visible: "language is everything, since it is the voice of no one, since it is the very voice of the things, the waves, and the forest" (VI, 155).

Let us consider the famous "Kanizsa triangle" (Fig. 4.1). If we describe only what occurs at the level of the "distal stimulus" (the physical stimulus) we would conclude that, ontologically speaking, the triangle does not exist. Yet we see it before us: it appears in perceptual space, superimposed upon another triangle and three black disks. We can observe it, appreciate it aesthetically, describe the slightly darker white within its perimeter, bounded by "quasi-phenomenal" contours; and we can share this perception with others. It is an intersubjectively observable and repeatable phenomenon.

To account for this appearance, this seeing of a triangle possessing unity and identity, we cannot dispense with phenomenological analysis. This analysis unfolds exclusively on the plane of direct experience. We perceive something meaningful, a *triangle*, and to account for this experience we must adopt phenomenology as our framework. Only at this level of complexity does the phenomenon exist and become

L. Taddio, *The Appearing of Sense*, Lecture Notes in Morphogenesis,
https://doi.org/10.1007/978-3-032-27021-4_4

Fig. 4.1 Kanizsa's triangle

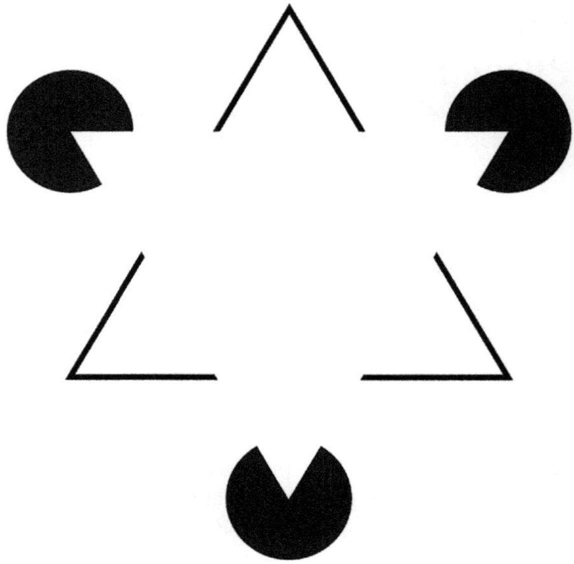

intelligible. Investigations confined to the distal stimulus (the physical level) or the proximal stimulus (the physiological level adjacent to the image-apparatus) cannot explain the phenomenon, for at those levels it is simply absent.

Is this an illusion? Does the "triangle" exist only because there is a subject to perceive it? Without a perceiver, neither the triangle nor the world would exist. Certain properties of reality emerge only under specific conditions, when particular relations are instituted. This is not a merely subjective perception, but one intrinsically bound to subjectivity in the transcendental sense. Phenomena emerge under given conditions, whether natural (a rainbow, a reflection) or artificial. The "technical" order remains continuous with the natural order, and both belong to the plane of experience. To make the triangle appear in a virtual environment as it does on paper, one would have to reproduce the same conditions that make its appearance possible in the real world.

We know that the white of the page is physically identical both inside and outside the figure, a fact easily verified by covering the black marks with one's hand. Yet this should not lead us to deny the triangle's existence. The essential task is to distinguish what we *know* about the object from what we *see*. We know that the physical stimulus is uniform, but we nonetheless *see* the triangle; conversely, we may fail to see things that are present at the level of distal stimulation, as in cases of camouflage or masking (PP, 21).

These considerations lead us to adopt what Köhler called the *stimulus error*: within a phenomenology of perception, we must distinguish what we know (about reality) from what we directly see (the world). The error consists in conflating these planes:

describing what we know instead of what we see. Making this distinction is straightforward when the difference between seeing and thinking is clear; yet, since no sharp boundary separates them, the closer they approach, the more the methodological application of the stimulus error becomes both complex and epistemologically fruitful.

The *stimulus error* must not be confused with the *experience error*, which consists in attributing to the stimulus the characteristics proper to experience. From an epistemological standpoint, employing both the stimulus error (PP, 9) and the experience error (PP, 5) allows us to approach the phenomenon correctly, maintaining a methodological distinction between the different perspectives of analysis: what we know about physical reality (the object of scientific inquiry) and what we directly experience (the object of phenomenology).

Let us now turn to the key notion of *immediate experience*, which underlies both Gestalt psychology and, indirectly, Merleau-Ponty's phenomenology. As an illustrative case, consider the "S–D psychophysical schema." This is the standard diagram used in experimental psychology or neuroscience to explain the causal mechanisms of perception. A first critique of this schema appears in Bertrand Russell; it was later elaborated and named by Paolo Bozzi to offer a visual representation of any possible causal explanation of perception.

Here is how it works. The schema traces the causal chain from object to perceiver. The blackboard becomes the very place where dualisms are inscribed: when we represent the world of experience on it, the marks "stand for" what we perceive. We are no longer living the experience in the first person, the only authentic sense of perception, but are instead confronted with an image, its "scientific" explanation. This is the classic laboratory situation: we are not ourselves perceiving, but observing someone else as they perceive.

The diagram unfolds from left to right: first the physical object (the distal stimulus), then its optical or geometric projection, then the proximal stimulus near the sensory apparatus, and finally we move to the retina (in the case of visual perception), to the neural processes within the brain. The directionality from stimulus (S) to detector (D) gives the schema its name: the "S–D psychophysical schema."

If we denote the subject's immediate experience by *phi*, where should we locate it in the schema? Intuitively, one might place it near the brain, as a mental correlate of neural activity. Yet this is already problematic. The schema encompasses various scientific domains: physics, chemistry, physiology, and neuroscience. But nowhere within it is there a proper place for first-person experience. It represents only the transphenomenal chain of events *underlying* experience.

To make phenomenology of perception an object of inquiry, we must first learn to distinguish between first-person and third-person perspectives, that is, between science and lived experience. Ernst Mach, and before him Goethe, had already pointed out that experience as such is not falsifiable: the act of seeing "here and now" a red spot on the wall cannot itself be falsified. Even if it were a hallucination, it would still remain an experience for us, just as in cases of illusion or perceptual ambiguity.

We see a man in the distance; approaching, we discover that "in reality" it is a scarecrow. We then say: it appeared to be a man, whereas in truth it was a scarecrow. What is falsified here is not the experience itself but the judgment "I see a man" and the description attached to it. Moreover, by changing our vantage point, we alter the very conditions of observation: returning to our original position, that same dark shape could again appear either as a man or as a scarecrow.

If this is granted, we can grasp the asymmetric relation between first-person and third-person perspectives. A new element discovered within experience may overturn entire explanatory theories, yet no fact established within the S–D schema can falsify first-person experience. "No explanation," writes Köhler, "can change a phenomenon or its location."[1]

Within *Gestalt* psychology, Koffka reiterates the same point when he states that "experimenting and observing must go hand in hand. A good description of a phenomenon may by itself rule out a number of theories."[2] The works of Köhler and Koffka were indispensable for Merleau-Ponty in the elaboration of his phenomenology: he fully grasped the innovative scope of the "new psychology". His engagement with the texts of the *Gestalt* authors would accompany him throughout his life, leading him to a profound excavation of this extensive scientific literature, including a close examination of the theoretical implications of its experimental component.

From within the Gestalt framework, Merleau-Ponty critiques certain tendencies, particularly the "objectification" of perception grounded in a causal model that treats the body as a merely physical system subjected to stimulation. Such models, he argues, can only fail (PP, 12): "Science has not and never will have, by its nature, the same significance *qua* form of being as the world which we perceive, for the simple reason that it is a rationale or explanation of that world" (PP, ix).

These remarks echoing our earlier reflection on the S–D schema already constitute, albeit implicitly, an answer to the question of whether phenomenology can be naturalised. In the end, it is another way of asking whether philosophy retains its distinctive role in relation to the sciences, or whether science instead exhausts the entire field of meaningful inquiry, whether, as Wittgenstein suggested in the *Tractatus*, everything that can be said meaningfully (that is, truly) belongs to the domain of the natural sciences (cf. 4.11).

---

[1] W. Köhler, *Dynamics in Psychology*, New York: Liveright, 1940, p. 71.

[2] K. Koffka, *Principles of Gestalt Psychology*, New York: Harcourt, 1935, p. 73.

Chapter 5
Seeing, Observing, and Explaining

In a 1952 letter to his friend, the philosopher and mathematician Maurice Solovine, Albert Einstein sketched a famous diagram illustrating the role of theory in relation to the plane of observation.

Figure 5.1 illustrates the process leading to the formulation of a scientific hypothesis (*System der Axiome*). It begins with our experience of the empirical world (*Mannigfaltigkeit der unmittelbaren – Sinnes – Erlebnisse*), in which events are observed. From this experiential plane, a set of hypotheses is then extrapolated, from which we deduce consequences (*gefolgerte Sätze*). These are ultimately compared with the initial level (the world of observables) in order to verify the validity of the deductions. In summary, the letter "E" designates immediate experience, while "A" denotes the axiomatic system from which conclusions are derived.

The curved line linking "E" to "A" does not originate directly from the horizontal axis but rises slightly above it. This displacement suggests that scientific knowledge, though rooted in the phenomenal world, can reach the plane of hypotheses only through a creative leap. Sensory experience provides meaning and content to "concepts" and "propositions" through its connection with them. "The connection of the latter with the former," Einstein observes, "is purely intuitive, not itself of a logical nature."[1]

We may regard the "A" level as resting, psychologically, upon the "E" level; yet for Einstein the transition from "E" to "A" is not logically grounded. It is an intuitive, psychological relation linking the two planes. From "A," by contrast, one may derive, through deduction, statements ($S, S', S'' \ldots$) that possess a potential claim to truth, a claim validated only by the return from "S" to "E," that is, by verification through direct experience. Even this final step, however, remains intuitive rather than strictly logical, for it involves a relation between conceptual inference and immediate experience. Einstein's diagram thus represents the scientific process as

[1] A. Einstein, *Autobiographical Notes*, Chicago: Open Court, 1979, p. 11. See also A. Einstein, *Letters to Solovine*, New York: Philosophical Library, 1987 and F. Laudisa, *Albert Einstein e l'immagine scientifica del mondo*, Roma: Carocci, 2015, Chap. 2.

L. Taddio, *The Appearing of Sense*, Lecture Notes in Morphogenesis,
https://doi.org/10.1007/978-3-032-27021-4_5

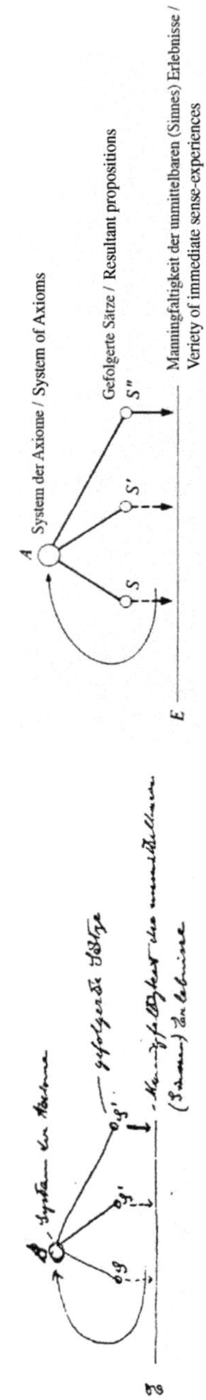

Fig. 5.1 Image of Einstein's scheme

an interplay between observation, hypothesis formation, and empirical testing. Of the many observations one could draw from this model, what interests us here is the role of observation itself. Far from being irredeemably "theory-laden," as is often claimed in the philosophy of science, observation remains the generative plane of knowledge, the level upon which theories find both their origin and their verification.

It is precisely upon this *beginning*, this plane of phenomenal experience, that Merleau-Ponty's phenomenological reflection is focused. He devotes a few brief yet significant remarks to Einstein (S, 193–194) and, in his Collège de France lectures on *Nature*, explores the meaning of the "thing" in light of contemporary physics. His aim is not to discredit science, but to question the gap that prevents Einstein from connecting the curved line of theory back to the plane of lived experience. It is upon this very ground that phenomenology situates its inquiry. For Merleau-Ponty, the entire edifice of scientific knowledge is rooted in the *lived world*. To think and to understand science rigorously means "a precise assessment of its meaning and scope", and to do so we must "begin by reawakening the basic experience of the world", of which science is but a "second-order expression" (PP, ix).

However much one may wish to emphasize the plane of immediate phenomenal experience as an autonomous domain of inquiry and as the proper object of phenomenological reflection, the question of its possible naturalization remains open. Traditionally, phenomenology has rejected this possibility; indeed, it was founded and developed upon this very assumption. Nevertheless, after Husserl, several attempts, some of them quite recent, have initiated a partial process of phenomenology's naturalization, ranging from developments in neuroscience to the proposal of a neurophenomenology. To naturalize phenomenology means to reduce perception, along with its meaning and its understanding, to the S–D schema, while acknowledging that its sense and conceptual grammar originate in direct experience.

In §17 of the *Monadology*, Leibniz presents a celebrated thought experiment intended to clarify the nature of perception within an anti-mechanistic (or, in contemporary terms, anti-reductionist) framework. "Perception and that which depends on it," he writes, "cannot be explained mechanically." To support this claim, he imagines a machine "whose structure makes it think, feel, and have perception." In effect, Leibniz is describing consciousness, the mind, and perception itself. The experiment continues by envisaging the machine as so large that one could enter it "as one does with a mill." To update Leibniz's metaphor, we might imagine, rather than enlarging the machine, shrinking ourselves down to enter, via a micro-capsule, the interior of the brain. Looking outward from such a capsule, we would see anatomical structures but nothing that could "explain a perception" (§17).

From a mechanistic standpoint, reality and the human body function like a complex machine: as in the case of a clock, complete understanding would consist in knowing every part and the causal relations among them. It is precisely this image that Leibniz seeks to overturn. Observing the *mechanisms* of perception—that is, the brain's physiology—does not amount to understanding perception itself. The same applies to the capsule analogy: however detailed our anatomical exploration is, we would never find anything corresponding to the *meaning* we ascribe to perception as lived. In contemporary terms, such exploration would place us within a single segment of the S–D schema: by studying the "mechanism," we would examine the presumed causes of perception, not the phenomenon of perception itself. From this perspective, Merleau-Ponty's remark takes on its full force: "Science has not and never will have, by its nature, the same significance *qua* form of being as the world which we perceive, for the simple reason that it is a rationale or explanation of that world" (PP, ix). Understood in this way, perception is the key to subjectivity, to our mode of inhabiting the world.

For Merleau-Ponty, the appearing of a thing to consciousness is a real operation performed upon the body of the perceiving subject; it cannot be described as if it were simply "the concrete spectacle of the world" (SB, 144). Direct experience concerns, strictly speaking, the totality of observables, not the transphenomenal causal reality underlying them. To claim that the meaning of perception derives from processes occurring within the organism is to remain within the S–D schema, that is, to reduce perception to a purely "causal" explanation, thereby transforming it into the object of investigation of a particular scientific field. From such a standpoint, the body does not enter perception as lived but appears as a transphenomenal entity. The notion of "inside" may refer either to a physical fact not directly observable or to an experience localised within us (for instance, a headache). We are able to situate things "out there" only by implicitly taking our own body as an "absolute here," the tacit frame of reference for our perceptual judgments (above, below, high, low, small, large, and so forth) within the field of phenomenal space.

An equally significant thought experiment is proposed by Wittgenstein in his *Remarks on the Philosophy of Psychology*. The example he offers resonates with Merleau-Ponty's claim that "science has not and never will have, by its nature, the same significance *qua* form of being as the world we perceive". Imagine, Wittgenstein suggests (his example was subsequently taken up and adapted by Bozzi), a neuroscientist who connects a subject's brain activity to a computer in order to record, in real time, the neural processes taking place while the subject observes a bistable figure such as the Necker cube (Fig. 5.2).

At a certain moment, the subject perceives the figure as "changing aspect," shifting its orientation. The scientist, noticing a corresponding change in neural activity on the monitor, is inclined to treat this as proof that "we see the figure differently." Yet Wittgenstein asks: "But with what right? How can the experiment make any pronouncement upon the nature of immediate experience? – It puts it in a particular class of phenomena" (§11). In other words, the neuroscientist, operating within the psychophysical S–D framework, cannot demonstrate *what* we actually see. The representation of experience on the monitor would serve as proof of what, precisely?

Such experiments reveal correlations between specific neural activations and direct experience, but not the content or quality of what is lived. When the graph fluctuates as the object's appearance shifts, in what sense does this "prove" what we experience at the phenomenological level? To know something implies the possibility of error or falsification; yet this possibility vanishes when we refer to the plane of direct experience. The experiment's value lies in constructing explanatory theories of brain function, not in capturing the lived meaning of perception. Phenomenal experience—what we see and live—may in principle arise from diverse causal chains

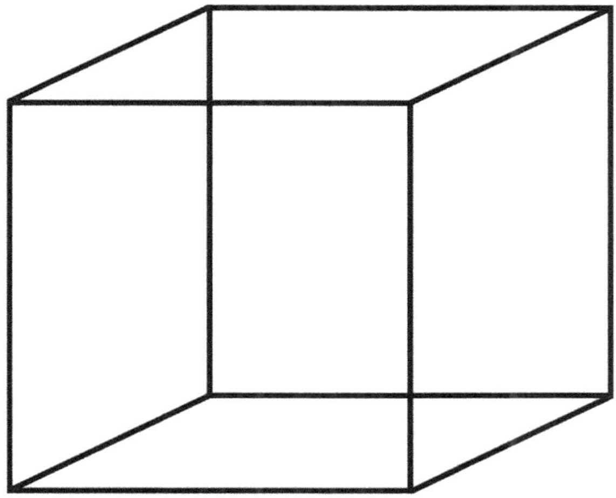

Fig. 5.2 Necker cube

leading to the same phenomenally experienced event. We might stimulate the eye directly, the brain itself, or produce equivalent perceptual surrogates, as in *trompe-l'œil* or virtual reality. Of course, the medium or substrate through which stimulation occurs is not irrelevant to qualitative outcome: the same shade of red applied to different materials will not yield the same appearance.

Like Merleau-Ponty, Wittgenstein, in his *Remarks on the Philosophy of Psychology*, engages critically with the Gestalt tradition and, more broadly, with the structure of direct experience, an inquiry conducted through the analysis of language and its conceptual grammar. Although Wittgenstein cannot be placed within the phenomenological lineage, he shares with it a rare aptitude for attending to phenomena and drawing from them subtle theoretical insight. These two thinkers, distant in formation yet convergent in sensibility, can be fruitfully compared through their analyses of perceptual-psychology experiments. Both recognise in these experiments a privileged terrain for clarifying the role and nature of perception in relation to higher-order cognitive activities such as language and thought. These ostensible and inter-observable dimensions of the world constitute the methodological foundation of a phenomenology of perception. Merleau-Ponty, however, goes beyond this level, employing the materials of experimental psychology to extract both epistemological and ontological implications.

We can, therefore, intertwine Merleau-Ponty's and Wittgenstein's perspectives by comparing their critical observations on the same perceptual situations, events, and examples. By maintaining constant the parameters of what is observed, shared observation allows for a comparison that transcends the purely conceptual level. Let us imagine, then, a laboratory of psychology of perception, where our two philosophers sit at the same table, joined by other observers. Together they describe, listen, and confront their interpretations, an exercise in inter-observation whose goal is not to clarify the perceptual process as such, but to bring to light the very features of the perceived object, guiding descriptions toward a formulation that best fits the phenomenon itself.

Consider now an experiment analogous to Wittgenstein's, following in the wake of Benjamin Libet's celebrated studies and carried out through functional magnetic resonance imaging. Before us lies another well-known bistable figure, Rubin's vase (1915), which can also be seen as two opposing profiles. By monitoring brain activity, researchers seek to determine which the subject perceives first: the vase or the faces. In a significant number of cases, activation of the right temporo-occipital tract, responsible for facial recognition, precedes conscious awareness by several seconds. Thus, even before the subject reports seeing the faces, the experimenter already "knows" it. In patients with lesions in this area, the opposite occurs: they perceive the vase before the profiles. Such experiments are fascinating, not only because they predict perceptual outcomes, but because they challenge deeply rooted beliefs regarding free will, though not necessarily to the point of denying it.[2] Yet even here, an essential phenomenological dimension is overlooked.

Recasting the experiment in Wittgenstein's terms, we could say that when the neural graph changes, the subject indeed perceives something different; yet this does not tell us what or how they perceive. The meaning of the figure (its shift of aspect) is apprehended by both experimenter and subject directly from the world, not from the monitor. The figure possesses its own expressive features: the impossibility of seeing both aspects simultaneously (as in the Necker cube or the Rubin's vase), or the distinctive emergence of form within a tenth of a second of observation. None of these qualities can be deduced from the monitor's data. The figure belongs to the order of meaning the world holds for us, grounded in the embodied system-environment in which we are situated (incarnate). However precise the experiment may be, it cannot capture the perceptual-expressive quality of the phenomena under investigation.

What is established by observing the monitor holds heuristic value for understanding the brain, particularly when considered in relation to direct experience. Yet what occurs in the brain does not exhaust the meaning of the "change of aspect", nor of what phenomenology can teach us about perception. The sense of what we experience finds no full correspondence in neural events at any point within the S–D schema. Appearance emerges from the process as a whole, from the dynamic totality of relations that constitute the S–D chain, and not from any of its isolated components, however precisely measured or technologically replicated.

Let us once again turn to the Necker cube. It "changes its appearance according as it is seen from one side and from above or from the other and from below. Even if I *know* that it can be seen in two ways, the figure in fact refuses to change its structure and my knowledge must await its intuitive realization. Here again one ought to conclude that judging is not perceiving" (PP, 40). Merleau-Ponty invites us to observe how, at times, we can deliberately provoke the restructuring of the cube; at other times it resists our efforts; and at still others it changes in spite of our will or intention.

---

[2] Cf. A. Benini, *Neurobiologia della volontà*, Milano: Cortina, 2022, p. 82.

Let us imagine being with another person who is unable to perceive the bistability of the figure. How could we determine which of the two cubes they are looking at? They might point to a set of lines, but these belong equally to both perceptual configurations. We are confronted with the same distal and proximal stimuli, yet with two distinct experiences, each accompanied by a different meaning. In one case, the cube occupies a certain spatial orientation; in the other, a different one. This prompts us to wonder whether it is indeed the same cube seen differently, or another altogether, positioned otherwise. At this stage, we would have to describe which face the subject perceives as foreground and which recedes into the background. But how can we be certain that they are truly perceiving the alternate configuration, that they have seen the restructuring of the figure, and are not merely asserting so without actually apprehending what they claim to see?

When the perceiving subject discovers the "virtuality" of the object, their body itself manifests a distinct bodily reaction attesting to the lived transformation within the field of experience. Of course, one may always feign or deceive; yet we shall set this aside, for the possibility of deception accompanies every report of another's perception. Indeed, it is itself a phenomenological sign of embodiment: deception, no less than sincerity, presupposes our being incarnate in a shared world, endowed with the capacity to anticipate, interpret, and respond to the gestures of others.

Chapter 6
Experimental Phenomenology

A separate chapter could well be devoted to the development of a project for an *experimental phenomenology*. At first glance, the very notion of naturalising phenomenology may appear oxymoronic, since, with Husserl, phenomenology was born and developed precisely on the opposite premise: the irreducibility of conscious life. Yet *experimental phenomenology*, understood as a "science of observables in act", can be conceived as a distinct and legitimate form of naturalisation. This approach is due above all to the work of Paolo Bozzi, who provided its most rigorous epistemological formulation by making explicit the experimental method of his teacher, Gaetano Kanizsa. Bozzi may be considered one of the last heirs of Gestalt psychology, a tradition that, in many respects, already constitutes a form of experimental phenomenology. Its aim is to determine the conditions under which a given phenomenon emerges, identifying the dependent and independent variables involved. The operations carried out for this purpose are strictly experimental yet remain entirely within the domain of lived observation. The entire range of operations thus stays co-planar with the field of the observable. This is the defining feature of experimental phenomenology: the element that distinguishes its method from other approaches in experimental psychology concerned with the study of perception.

To reflect philosophically upon the experimental dimension is to recognise, on the one hand, that experimentation, though indispensable, is not in itself sufficient to constitute a theory of knowledge; and, on the other, that philosophy cannot claim complete autonomy from scientific knowledge, since it must draw upon their results and methods. More broadly, philosophy may be understood as the critical reflection upon the scientific knowledge of its time. To comprehend the nature of perception and its implications, we must adopt a dual register, scientific and philosophical. Merleau-Ponty exemplifies this hybrid stance, situated between an ante litteram experimental phenomenology and the Husserlian phenomenological investigation.

Doing philosophy therefore entails a preliminary reflection on the presuppositions underlying the very problems one intends to confront. For this reason, it is essential to place at the centre of inquiry the analysis of *immediate experience*, a key

L. Taddio, *The Appearing of Sense*, Lecture Notes in Morphogenesis,
https://doi.org/10.1007/978-3-032-27021-4_6

concept shared by both the Gestalt and phenomenological traditions. Merleau-Ponty perceives the reciprocity between philosophy and science: from Gestalt theory he inherits the notion of "immediacy"; from Husserl, the concept of *epoché*, the analysis of the implicit mediation at work in every cognitive process.

If experimental psychologists of perception are charged with extending the scope of experimentation, philosophy—and phenomenology in particular—has the task, as Merleau-Ponty did, of extracting from it epistemological and ontological insights that allow for a broader understanding of the lifeworld: the clarification of the meaning that relations disclose and that the world reveals. Our natural perspectivism, as a part that reflects upon the whole from within, generates the familiar "dualisms" that oppose a presumed autonomous individuality to the surrounding world. Naïve realism conceives world and subjectivity as independent entities rather than as correlative terms of a relation. For Merleau-Ponty, the difficulties encountered by realism "arise precisely from having tried to convert this original relation into a causal action and to integrate perception into nature" (SB, 193). In so doing, realism is equated with "external reality" understood in a purely physical sense. Merleau-Ponty criticises this implicit assumption and its lack of analysis and justification, a critique he extends to the sciences as a whole. Naïve realism consists in accepting the independence of the world with respect to the experience we have of it. Yet this "dependence" belongs to first-person observation: from an ontological standpoint, world and lived body are one and the same, inseparable, since they form a single structure. If we are still to speak of realism, it must be a realism of *relation*, a realism of the process itself.

The constitutive intertwining (*chiasm*) between world and corporeality (*Leib*) entails a constant problematisation of the ontological in relation to the epistemological, and vice versa. The meaning of this relation may be grasped through a careful phenomenological analysis and, at the same time, investigated scientifically at the ontological level; or it may first be approached through scientific inquiry and subsequently extended phenomenologically to the lifeworld. To apprehend form (*Gestalt*) in its genesis is to recognise subjectivity and world as internally related dimensions of reality, both at the physical and the phenomenological level, inasmuch as reality itself can be described in relational terms. Phenomenological and scientific knowledge, epistemological and ontological reflection, are thus interwoven in our very conception of the world as one of the possible planes of Being within reality. The world, understood as a plane of emergence of reality, is constrained rather than arbitrary: each form of life embodies a specific level of internal relation to reality.

The meaning we discover in action is an expression of our form of life, rooted in the relative stability (or metastability) of a world that gathers our gestures, generating new ones in turn and producing further transformations and information. The primary or *proto-value* that things possess for us derives from this relative stability of the world. Were the world unstable, it could not sustain any coherent structure of behaviour: we act on the basis of the world, not of reality considered in abstraction. For Merleau-Ponty, there is no objective reality underlying the world, for this would entail absolutising one among the possible perspectives. It is precisely at this juncture that he directs his main critique at Gestalt psychology, insofar as it tends to relapse into a subtle objectification of the world.

Chapter 7
Beyond Gestalt: Structure and Meaning

Let us return to the example of Mach's square. Merleau-Ponty employs it to level a critique against Gestalt psychology, exposing residual commitments to realism: lingering assumptions of objectivity that persist despite its major theoretical advances. His aim is to underscore the insufficient analysis of the epistemological and ontological presuppositions underlying the psychology of form. In contrast to the Gestaltists' reading of Mach's square, which stresses the functioning of the visual field by highlighting the dissimilarity between the two figures, Merleau-Ponty observes that judgment, insofar as it involves asserting the equality or difference of the figures, presupposes retaining the first as a reference. We perceive the oblique figure as a square not because we immediately refer to the square as seen frontally, but "because the diamond-shaped appearance in an oblique presentation is immediately identical to the square appearance in a frontal presentation" (PP, 350). Yet these appearances are not *immediately* identical: we recognise them as such only subsequently, through a process of calculation or comparative reasoning applied to the figures. The two configurations are, in fact, immediately perceived as different: the oblique figure does not "remain a square" (PP, 350).

Through the various configurations, Merleau-Ponty claims, we are given "that orientation of the object which makes it possible, and because these shapes are presented in a context of relations which, a priori, equate the various perspective presentations with each other" (PP, 350). However, his description does not fully correspond to what is phenomenally experienced, since it follows the thread of reasoning rather than observation. He insists upon "equivalences" in order to eliminate the presumed reference to an "objective" square as the standard against which the oblique figure (the diamond) is measured. In this way, the perceptual judgment would be grounded in the objective status of the first figure, which remains the reference parameter. Yet the second figure is perceived as a diamond not by virtue of *a priori* equivalences, but on the basis of the concrete experience of the relational field: it is the object's orientation that establishes a direction of reading and a distinct phenomenological grammar, transforming the square into a diamond. This

L. Taddio, *The Appearing of Sense*, Lecture Notes in Morphogenesis,
https://doi.org/10.1007/978-3-032-27021-4_7

misreading can be explained only if we take into account Merleau-Ponty's deeper intent. In his view, Gestalt analysis continues to presuppose a background of objectivity, a form of scientific realism. By contrast, Merleau-Ponty seeks to return to a primordial level of perception, anterior to any comparison between the first and second figure and prior to any judgment—including the one that would lead us to say that the second merely "appears" as a diamond while "in reality" it remains the same square. Alternatively, following the Gestaltists, one might claim that it is a *different* figure—but on what basis? The example, rather, allows us to compare two appearances, the square and the diamond, without attributing greater reality to one over the other. Perhaps Merleau-Ponty's deeper suggestion is that any judgment about the relation between appearance and reality already presupposes an epistemological position regarding the domain of appearing itself. In this sense, the identity or difference of the two figures presupposes two distinct frames of reference—one geometric, the other phenomenological—that is, two epistemological perspectives grounded ontologically in the field of appearance.

Through what is, strictly speaking, an inaccurate description, since it constitutes a *stimulus error* to treat the two figures as perceptually identical merely because we know them to be geometrically so, Merleau-Ponty nevertheless affirms an essential insight: there is no "appearance" ontologically opposed to "reality," but rather coplanar phenomena, distinguishable and reversible according to the critically adopted frame of reference. The first figure does not serve as an absolute standard for the second, for the latter is not "apparent" in relation to the "reality" of the former. Merleau-Ponty compares this to the case of the cube, where, despite "sides deformed by perspective," the object is nonetheless perceived as having six equal faces. It is precisely through that deformation, judged as such only when isolated from context in an empiricist framework, that we apprehend the equality of the cube as a thing (PP, 350). We have already noted that Merleau-Ponty's thought may be read as a critical footnote to Husserl: if Husserl conceives the *lifeworld*, Merleau-Ponty furnishes its phenomenological counterpart in the "return to the phenomena" he enacts through his phenomenology of perception. The theoretical distance between them lies in the interpretation of the implicit mediation within the immediacy of perception. Conversely, Merleau-Ponty may also be read in a critical relation to Gestalt psychology: drawing on that tradition, he recovers the concept of *immediacy* and integrates it into phenomenological reflection. In doing so, he weaves together the Husserlian and Gestalt paths, opening them to new terrains of inquiry and to previously uncharted research perspectives.

Merleau-Ponty's intellectual horizon is remarkably broad. He engages in sustained philosophical dialogue not only with Descartes, and with the enduring reverberations of dualism in Western thought, but also with Hegel, Nietzsche, Marx, and Freud, among others. His conversation extends to Bergson and, more directly, to Sartre, as well as to pedagogues and anthropologists such as Lévi-Strauss. Yet his meticulous engagement with the writings of Gestalt psychology remains indispensable for understanding his method. This is particularly evident in his posthumously published lecture courses, especially the one delivered at the Collège de France in 1953, where he analyses experiments by Wertheimer, Koffka, Köhler, Guillaume,

Lewin, and Michotte. Merleau-Ponty's interest lies in their methodological procedures and theoretical presuppositions, which he contrasts with the empiricist and associationist psychologies that preceded them. Beyond the notion of form, the true nucleus of this legacy is the concept of *immediate experience*, which runs through the entire Gestalt tradition and its antecedents, from Mach to William James.

However reworked, the concept of *Gestalt* remains foundational to Merleau-Ponty's reflections. One might even ask how much of his philosophy would remain if we were to remove the elements inherited from Gestalt theory and from the sciences. Yet this does not imply that his thought can be reduced to the statements of the Gestaltists. The difference lies in what we might call the "theoretical residue," that distinctive oscillation between an internal and an external stance toward the sciences that defines his philosophical approach and points beyond it. The sciences, psychology, and the epistemology of Gestalt provide him with the material upon which to exercise and test his reflection. Just as experimental science is indispensable for investigating reality, phenomenology is equally indispensable for critically examining subjective experience and the dimensions of meaning inscribed within the world.

Merleau-Ponty's analysis of the structure of behaviour led him to adopt the notion of form, derived from Gestalt psychology, as a conceptual instrument for analysing both behaviour and perception. Behaviour can be understood only if grasped in its overall meaning in relation to the world-environment (*Umwelt*). This holistic vision is intimately tied to the notion of form. The form of our actions cannot be reduced to an analysis of their isolated components, whether purely material or purely psychic, but must be apprehended as a structure. The unity of meaning in conduct cannot be reduced to an "internal" analysis of the subject, to the quantitative laws of the physical world, or to a simple "stimulus–response" relation. Form possesses its own characteristics, internal laws of organisation, and values that, in ecological terms, function as indices of action for the body in its relation to the environment. The body lives, acts, and transforms its milieu; and the sense of its action cannot be superimposed upon a mere causal chain.

Action thus involves a level of complexity requiring conceptual categories adequate to the analysis of behaviour, a narrative space wherein our form of life becomes manifest. When understood at this level, perception reveals the meaning disclosed by the relational field itself. The world-environment obeys a condition of relative stability, its structural correlation with the lived body, which allows for coherent forms of conduct. This implies recognising not only the dynamic processes within the perceptual field, but also a circularity that must be understood according to systemic logic: we act through open, dynamic behaviours, not via linear stimulus–response mechanisms. Perception does not merely receive information; it processes, transforms, and shares it in an intersubjective dimension, recoding it through kinaesthetic patterns of action. The sensorimotor system, in fact, is active from the earliest stages of development.

In continuity with Gestalt theory, Merleau-Ponty challenges the identification of seeing with thinking: those who see and those who are seen belong to a dynamic system governed by "principles of reversibility." These feedback-driven, circular processes weave the subject into the world, allowing us to grasp how, on the one hand, the body is predisposed in relation to its environment, and on the other, how the environment itself shapes and selects the inputs to which the body responds. Approaches that polarise analysis toward either the subject or the object neglect this circularity, the fact that our behaviour does not follow "mechanical" patterns. We are not automata, and our responses to the environment are not rigid but flexible, adaptive, and dynamic. Feedback and reversibility bind us to our surroundings, such that meaning emerges and stabilises within our actions. An excessively uncertain or indeterminate environment would preclude this stabilisation of meaning: we perceive a table or a tree, each a structured form imbued with significance, rather than a swarm of atoms in ceaseless motion.

For Merleau-Ponty, the limitation of Gestalt psychology lies in its failure to generate a genuine "set of new categories" (PP, 56). This absence of critical elaboration prevents the Gestaltists from escaping, at least implicitly, an objectivist mode of thought. To take this further step, Merleau-Ponty reinterprets and reactivates the Husserlian project by bringing it into dialogue with the "new psychology." In so doing, he applies what he learns from Gestalt theory to Husserl, above all, the notion of "immediate experience" beyond its experimental meaning, and, conversely, brings to Gestalt psychology what he inherits from Husserl regarding the transcendental and phenomenological dimensions of inquiry. This double displacement generates a multilayered discourse that still leaves open many theoretical questions concerning the compatibility between the transcendental approach and experimental psychology, as well as the possibility of a naturalised phenomenology. Relative to Husserl, Merleau-Ponty recalibrates and reinterprets the role of intentionality and of the *epoché*, thereby distancing himself from the framework of *Ideas I*. At the same time, he departs from Gestalt psychology, criticising it for remaining captive to the "evidences" of science and for reinstating, albeit implicitly, an unexamined realism with regard to the world. It must nonetheless be acknowledged that *Gestalttheorie* itself claims a phenomenological dimension, though not one directly derived from Husserl. Reading Merleau-Ponty today can help revitalise this reciprocal dialogue. At these intersections, one cannot overlook the theoretical ambiguities and oscillations that must be re-examined beginning with the concept of *epoché*, without which, in one interpretation, phenomenological reflection could not even begin. Alternatively, one might regard the *epoché* not as indispensable but as one significant gesture among others, a necessary pause, yet not the only possible point of departure for phenomenological investigation.

Chapter 8
Epoché, Ontology, and the Emergence of Being

Whether or not the epoché is the *conditio sine qua non* of phenomenological practice, one may still ask: where do we locate the point of departure for our relation to a world grasped in its becoming? Since an absolute or abstract thematization of the beginning is impossible, we must start from the contingent situation of the life-world in which we are here and now incarnate. This situation is intrinsically complex (structural, field-dynamic, and historical) implicating both world and subjectivity. In the legacy Merleau-Ponty inherits from Gestalt psychology, he finds a strategy for reflecting upon and reinterpreting Husserl, the very notion of the *epoché*, and the "enigma of the transcendental."

Husserl's *Analyses Concerning Passive Synthesis* suggest that the epoché is not the sole key to phenomenological inquiry, even if it bears decisive weight by opening the field of "reduction" and, consequently, of the transcendental ego (cf. PP, 57). The suspension of judgement regarding the existence of the external world (*epoché*) has often been absolutized by interpreters, treated as a necessary condition for accessing transcendental consciousness. In Merleau-Ponty the matter is more nuanced and, at points, contentious: descriptive inquiry tethered to immediacy takes precedence over the "reconstructive" logic of constitution that characterizes Husserl's project.

The chief lesson of reduction, Merleau-Ponty insists, is the "impossibility of a complete reduction," and yet to rediscover the world we must "break with our familiar acceptance of it" (PP, xv). This is not a single method but a set of repeatable practices. The philosopher is an "eternal beginner," continually calling into question the grounds of thought, remaining close to the point of departure, and cultivating a critical-phenomenological vigilance toward the beginning of thought itself. Recovering the "original wonder" entails a twofold movement: a "loosening" from the world as merely usable within the natural attitude, and an "approach" grounded in an eidetic grasp of essences.

Merleau-Ponty situates contact with the world at the crossroads of Husserl and Gestalt. Within an anti-foundational framework, against the search for a final, definitive ground, he, with Husserl, problematizes the beginning of knowledge: the *epoché*

L. Taddio, *The Appearing of Sense*, Lecture Notes in Morphogenesis,
https://doi.org/10.1007/978-3-032-27021-4_8

as an articulation of grounding that is never univocal but always situated, each time, within a concrete existence. In this sense, for both thinkers, it is the very condition for undertaking a phenomenological and transcendental analysis. This entails bracketing the "natural world" in view of the reduction that leads from the empirical ego to the "transcendental ego," together with a radical shift of attitude: not practical absorption, as in the natural attitude, but an orientation to the essential (invariant) features of the life-world on which Husserl seeks to shed light.

Thus understood, the *epoché* articulates a radical problematization of the beginning as an elucidation of the process implicit in immediacy, in relation to what is immediately present to consciousness. It is precisely at this intersection between immediacy and the clarification of immediate content that the relation between Husserl and Merleau-Ponty is decided. The *epoché* localizes the beginning, our "here and now" as historically, culturally, and intersubjectively determined. Whatever lies within this contingent space becomes the object of reflection: our contact with the world before a determinate thing, within a determinate *milieu*, brings forth the implicit stance named by the *epoché*. Because this condition is contingent, it cannot be construed as an "absolute"; yet what appears in immediacy becomes an absolute object of inquiry, and the nexuses that constitute this complex experience are available to phenomenological analysis. In this sense one can speak of a "genetic phenomenology" as well as a "static" one, which investigates the constitution of perceptual content, and of a "constructive phenomenology" (cf. PP, viii) addressing the implications of the lived body. This (pre-scientific) field embraces the full array of aspects involved in perception, carrying out a transcendental reflection upon the intersubjective structure of consciousness. Both alterity and our body are born, as it were, from an "original ecstasy" (cf. S, 174).

What is made evident thereby becomes a locus of foundation capable of accommodating the "paradox" of its constitution within an anti-foundational epistemology. From this perspective, Husserl, through the reduction consequent upon the *epoché*, both overcomes the natural attitude and preserves "the whole world of the natural attitude" (S, 162). If, on the one hand, Merleau-Ponty can say that phenomenology studies essences (and the operations that render them accessible), on the other hand it is also the method that returns essences to existence, enabling us to grasp the meaning of human beings and world (PP, vii ff.). This double register—affirming the essence of the thing while apprehending its here-and-now manifestation through its adumbrations—characterizes phenomenology.

From Husserl, Merleau-Ponty inherits the "awakening" of the world: philosophy as transcendental interrogation of the natural attitude and, in a reciprocal movement, as evocation of the world "already there," prior to any assertion, concept, or word. The task is to question philosophically that original, naïve contact in order to confer upon it a "philosophical status" (PP, vii). In the Preface to the *Phenomenology of Perception* he sketches this program, which finds full development across the work; yet to understand his intent the Preface must be read alongside the Introduction, where the path back to "the things themselves" is traced. With Merleau-Ponty it is a matter of relearning how to look, placing at the center the constitutive relations of subjectivity, in an exercise that engages how we embody the dynamics of meaning.

We may decenter the relation between Merleau-Ponty and Husserl by reactivating direct contact with the world through Gestalt experimentation. In this way the very idea of phenomenology is renewed in relation to our present world, opening inquiry to what recent psychology and, more broadly, the contemporary sciences have taught us.

Phenomenology does not ground knowledge on essences (*eidos*) as eternal, immutable models detached from contingency and becoming. Eidetic structures are bound to existence and thus to experience; hence talk of regional ontologies or domains of knowledge. Such investigations do not converge upon a transcendent point in the manner of classical metaphysics: transcendence arises within immanence. Responding to a logic of "how" rather than "why," what appears, appears as it appears; through *epoché*, understood as an "opening," one seeks the root, the stratifications of meaning, and the genetic analysis of the visible through the thing's operative intentionality, grasped in its possibility and actuality. This clarificatory work is not abstractly universal; it must be renewed each time for each category of subjects. It is a practice that must be constantly reactivated as an "opening toward." Merleau-Ponty receives this Husserlian legacy while lightening the theoretical weight of the *epoché*, so as not to bind original contact with the world to the transcendental ego (as in the eidetic reduction). He thus widens the transcendental register even as he preserves the practice of remaining near the world through direct experience. Remaining near the world (*epoché*) means discerning the sense that emerges from forms and from the "intersections" among multiple intersubjective experiences.

Chapter 9
Meaning, Existence, and Form of Life

"Because we are in the world, we are *condemned to meaning*" (PP, xxii): this statement reveals the existential depth Merleau-Ponty ascribes to phenomenology. We are bound to meaning insofar as we are "condemned" to have a world: an empirical, phenomenal plane of immanence in which we live through the network of relations instituted by the lived body (*Leib*, the lived body) as perceptual system. We never step outside the actuality of world and perception. Meaning, world, and body present themselves in correlation as a "structure": to analyse them is to seize the very flesh of their being, rather than dissecting them into separable parts. Even when we adopt a "neutral" scientific gaze—mathematical, quantitative, discrete, digital, inhuman—our comportment toward the world remains oriented by sense: qualitative rather than quantitative, continuous rather than discrete, analogical rather than digital. Even in using science and manipulating the world, we fall back into a horizon of meaning. Our existence is tied to an order of meaningfulness: "death" is not a worldly event in the strict sense, for we cannot embody or live it without contradiction; we experience it only analogically, through the other. We always find ourselves "already born" or "still alive," "thrown into a natural world," and in this horizon we can interpret "birth" and "death" only at a distance, by reflection, "only as prepersonal horizons" (PP, 250). Meaning "stands" there, dynamically sedimenting within our "field of existence" (the transcendental field), a notion that is reworked in Merleau-Ponty's later thought (VI, 171), where he sketches a new ontology of the flesh in the Working Notes to *The Visible and the Invisible*.

Just as the world is saturated with meaning and inseparable from it, so our adherence to the world is neither deliberative nor chosen. Grafted into the sensible, our body "thinks" according to its logic (preindividually, as Simondon would say, drawing on Merleau-Ponty). We do not determine our own incarnation, just as we have no direct consciousness of our birth or our death, two horizons foreign to the world that lead toward the primordial generativity of Being. The world's forms (*Gestalten*) are metastable, in continual transformation (visible/invisible): the collapse point of a form of life is what we interpret as "death." In the depths of Being there are no

L. Taddio, *The Appearing of Sense*, Lecture Notes in Morphogenesis,
https://doi.org/10.1007/978-3-032-27021-4_9

stable essences, only the reiterated genesis of formative processes; the elemental registers (air, water, fire, earth) must be conceived neither as substances nor as spirit, but as "flesh" (VI, 218). Our body, like every Gestalt, participates in this flesh: we are incarnate, woven into the fabric of Being, and this chiasm grounds our "carnal participation in its sense" (VI, 208). The horizontal planes of the world proper to each form of life intersect with the "verticality" of Being: such is the logic of the carnal universe that Merleau-Ponty sketches at the end of his work.

Always from within life and the world as horizon of meaning do we grasp both world and meaning in the "perceptual field", beyond which there is no "outside" or "beyond" that could be experienced. The constitutive "limits" of the field of existence, in the transcendental sense, are not restrictions on knowledge but the very condition of its possibility: they institute knowledge as an opening to the truth of Being. Without this dynamic field, we would not have the vibrations of space and time proper to incarnation, nor the relative stability of the world. The relation between world and Being mirrors that between the visible and the invisible (VI, 256). Analysing the visible entails positing the invisible, not as a merely "possible" visible, but as a distinct dimension sharing a common weave with the visible, namely flesh. Visibility itself implies a "non-visibility" (VI, 247). The visible is invisible by "surpassing": the invisible depth of Being generates the visibility of the world and our field of existence. In this opening to Being lies the truth of perception, our being-in-relation to the world and to the flesh of "wild" Being, enfolded in its verticality. "But from this it follows that the words most charged with philosophy are not necessarily those that contain what they say, but rather those that most energetically open upon Being, because they more closely convey the life of the whole and make our habitual evidences vibrate until they disjoin. Hence it is a question whether philosophy as reconquest of brute or wild being can be accomplished by the resources of the eloquent language, or whether it would not be necessary for philosophy to use language in a way that takes from it its power of immediate or direct signification in order to equal it with what it wishes all the same to say" (VI, 102–103).

There is a "thickness of original acquisition" (PP, 468) between us and the sensible that prevents our synchronization with the world from being transparent. It does not belong to deliberation; it is tacitly presupposed whenever we perceive. It is given "prepersonally," as a primordial contact with the world: this is what we meant by the word "meaning" (cf. PP, 170). Through perception we can only draw near to the beginning—always a particular situation within the becoming of world and things—just as with a work of art, which stands close to the mystery of its own genesis through color, form, value, its constitution as a "world." We cannot explain or grasp creation except in its unfolding, in its singular process, in the very betrayal of the first rupture of equilibrium when the artist begins.

When we look at a slashed canvas, a mark, a gesture, a nuance, we remain bound to the sense in which the world's transformations are given and sustained for us. To thematize the origin of forms is, paradoxically, to find oneself inevitably on this side of meaning, confronted with the impossibility of that first act. This epistemic impossibility returns us to ontology: there is no true "beginning," only ceaseless transformation. To grasp the stability of sense is to take a stance toward Being, dictated by bodily presence and by our belonging to a perspectival horizon, a form of life. The edges of the perceptual field are not fixed margins; they evoke the continuity of sense saturating the world. We perceive space as enveloping us, extending beyond our gaze, behind our backs. In this horizon, meaning appears as a necessary illusion consistent with our form of life.

Describing the perceptual field is never simple, for it entails a gap introduced by our body as epistemic stance in space: we perceive only partially; lateral boundaries blur into that *je ne sais quoi*, that "good vagueness" of an essentially indeterminate vision. At its sides, the field spills beyond the optical cone; there is no sharp perimeter. Continuity of surrounding space prevails: "what is behind my back is not without some element of visual presence" (PP, 6). Unlike many animals, our visual field spans roughly 180° horizontally and 140° vertically, its partial limitation is readily compensated by head rotation, what Gibson calls the "eyes-in-the-head-on-the-body-resting-on-the-ground" system. Perception does not arise from the sum of isolated senses, nor does it reside in parts: the body itself is a perceptual system. For Gibson, as for Merleau-Ponty, these are "systems with nested functions."

The visual world envelops us as though we were inside a sphere; it presupposes our capacity for movement, whereas a "field" presents itself as a configuration requiring head and eyes to remain still. "My body is wherever there is something to be done" (PP, 291). At each crucial juncture Merleau-Ponty shifts from the natural to the philosophical register, from the "phenomenal field" to the "transcendental field." This is to affirm the originality of the phenomenal life of the body and of the primordial world, prior to any attempt at objectification. Even the notion of Gestalt is reinterpreted: not as a mere condition of possibility, but as the very appearing, in the generative sense, of the world itself (PP, 70).

The process that leads us to acknowledge the primacy of the phenomenal field over its objectifications proceeds from the former toward the transcendental field (PP, 69). Hence "phenomenology is a phenomenology, that is, a study of the advent of being to consciousness, instead of presuming its possibility" (PP, 71). Philosophy seeks to elucidate the implicit transitions between passive and active, mediated and immediate, without "taking place in absolute consciousness" as the seat of absolute knowledge. Its task is to analyse the gradations by which consciousness is achieved: "radical philosophy" consists in this work of clarification.

Consciousness, then, is not a solution but a problem (cf. PP, 73). Understanding perception begins with the natural attitude and leads to the transcendental, not as a timeless condition of possibility, but–following Husserl–as a reference to the operative process of constitution. On the phenomenological side, this work of clarification can converge with Wittgenstein's clarification in the conceptual analysis of language: two trajectories that meet at a shared point. As Merleau-Ponty notes, "nothing is more difficult than to know precisely what we see" (PP, 67); and, in parallel, Wittgenstein: "How hard I find it to see what is right in front of my eyes".[1]

[1] L. Wittgenstein, *Culture and Value: Revised Edition*, Oxford: Blackwell, 1998, p. 44.

Chapter 10
The Phenomenal Field and Its Blind Spot

By the "immediate" we do not mean a fusion of subject and object, but rather—more precisely—the sense, structure, and spontaneous arrangement of parts (cf. PP, 67). On the basis of the stimulus error, we can distinguish physical reality from the phenomenal world as correlate of perception; likewise, we can distinguish what we directly perceive through our body from physiological aspects, i.e., unnoticed internal processes. For Merleau-Ponty, "the properties of the phenomenal field" call for natural language, not formal languages (logical–mathematical), which remain at a distance from the lived world; otherwise, "it becomes impossible to reconstitute the descriptive content of perception, the actual view of the world, as an effect" (SB, 193).

Let us take as our frame of reference the field of experience, a notion Merleau-Ponty inherits from Gestalt psychology and which, by analogy, recalls a decisive turn in late nineteenth-century physics. Just as physicists posited the electromagnetic field, an invisible network of interactions that explains visible phenomena, so psychology introduced the visual field to describe and explain, via self-organizing dynamics, what occurs within perceptual experience. Merleau-Ponty analyses the "phenomenal field" as immediate experience, in contrast both to causal scientific explanations and to the residual mechanism, empiricist associationism, and atomism that accompanied behaviorism and other psychologies of his time. We must first clarify "immediate": it does not refer to raw impressions, nor does it introduce a new empiricism; rather, in the field the "arrangement of parts", "sense", and "structure" are what is immediately given (PP, 67). The phenomenal field is not an "inner world", "state of consciousness", "psychic fact", "introspection", or "Bergsonian intuition" (PP, 66–68). It requires a holistic approach: an exact description of relational dynamics (irreducible to linear cause-effect relations), of the self-organization of unities, and of part-whole relations, even where these cannot be traced back to past experience. Our immediate experience may be captured by Wittgenstein's dictum: "And from nothing *in the field of sight* can it be concluded that it is seen from an eye" (5.633).

In his *Tractatus*, the following image appears (Fig. 10.1).

L. Taddio, *The Appearing of Sense*, Lecture Notes in Morphogenesis,
https://doi.org/10.1007/978-3-032-27021-4_10

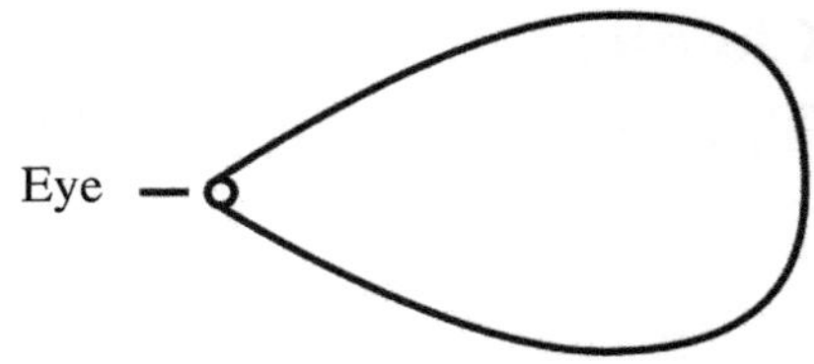

Fig. 10.1 Wittgenstein's drawing

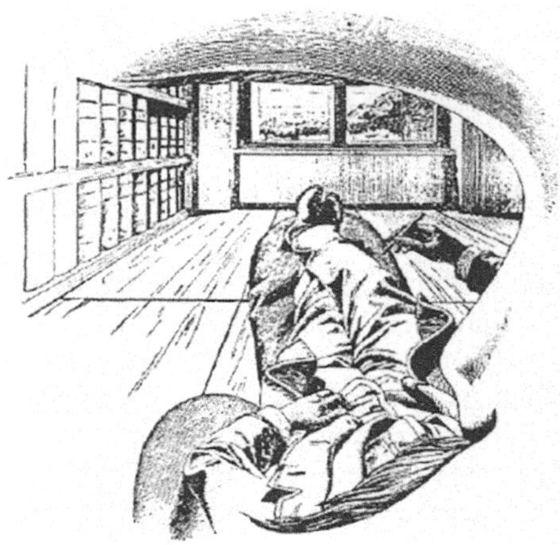

Fig. 10.2 Visualization of the perceptual field

The image drawn by Wittgenstein recalls the famous illustration (Fig. 10.2) reproduced by Mach in *The Analysis of Sensations*, a foundational text for Gestalt psychology. Both images clarify the "visual field": we see the world in the first person, with blurred contours and a portion of the body entering the scene; the field lacks a sharply defined perimeter. The effect is akin to peering through a keyhole: a static frame of observation, without movement. Precisely this immobility grounds Gibson's critique of Gestalt experimental setups within his ecological approach to perception. His objection concerns the laboratory's static subject and impoverished stimulus arrays, reductions designed to isolate invariants. For Gibson, invariants are anchored in the information of the optic array in an actual ecological context. In ordinary perception we move through richly structured environments; movement constantly furnishes new information. Building on his critique of "frozen" arrangements, Gibson develops the notion of the "visual world" (later integrated with affordances), defined through its anchoring in the head-on-the-body-resting-on-the-ground within an environment. These interlinked elements constitute a perceptual system that inherently includes possibilities for action and motion. This perspective converges with Merleau-Ponty's account of the body as perceptual system and as an indispensable component of any theory of perception.

With these clarifications, the point identified by Mach and systematically developed by the Gestaltists is the distinction between immediate, first-person experience (genuinely phenomenological) and third-person accounts, i.e., causal studies exemplified by the S–D schema. When treating perception we must not confuse what we know about reality with what we directly see of the world. Methodologically, the stimulus error is linked to a narrower sense of *epoché*: bracketing what we know about reality in order to attend to the world as it manifests itself. This requires a "step back" from our pragmatic, manipulative comportment, an attitude keyed to the expressive affordances of things. Such a methodological stance marks a point of contact between experimental phenomenology and Husserlian phenomenology. Guarding against the stimulus error allows perception to be grasped in its proper sense: direct experience of world and things. The same holds in experimental phenomenology, provided one understands "suspension" here as the temporary setting-aside of background knowledge about the world.

Wittgenstein's words render Mach's image verbally: nothing within the "visual field" licenses the inference that what is seen—the world—is seen by an eye. In his drawing, Wittgenstein marks the point where the eye is inscribed into the world, as if to indicate the condition of possibility for experience. At the root of empirical-phenomenological knowledge lies a constitutive opacity, a non-transparency from which experience originates. This limit-point does not belong to direct worldly experience but, physiologically, functions as the trigger that embodies vision and situates experience ecologically. The visual field is thus anchored in a *punctum caecum* at the base of consciousness: "*What* it does not see it does not see for reasons of principle, it is because it is consciousness that it does not see" (VI, 248). For Merleau-Ponty, the very fact of being conscious of the world and of its possibility implies the necessity of this blind spot: "*What* it does not see", he writes in a working note, "is what makes it see" (VI, 248).

There is, moreover, a sense in the world that is "returned" to us (reversibility): we look, and we are also objects of the other's gaze; in a certain sense, as many artists attest, "things look at us." To be both seeing and visible, exposed to the other, renders perception in part narcissistic (cf. VI, 139). This play of reflections and intertwinings (*chiasm*) among self, others, and world provides the shared fabric within which we can grasp the difference between eye and gaze, and, correlatively, the distinction between science and experience and the role phenomenology assigns itself: to apprehend the complexity of the phenomenal world, irreducible to a purely physical analysis of reality. Something is always given to consciousness that never fully coincides with lived experience, with "the thing itself": this "gap" founds the constitution of meaning and becomes central in Merleau-Ponty's late work at the intersection of psychoanalysis and phenomenology, hence his dialogue with Jacques Lacan (cf. *Seminar XI*).

In the final phase of his thought, perspectives on the "thing" converge in an ontology of multiple forms and virtualities emerging from one and the same being. It is from "my flesh," from "my body," that "the cube itself"—the thing itself—"closes the circuit and completes my own being-seen" (VI, 202). Being appears as the encompassing of ourselves and things. Merleau-Ponty seeks to thematize process as such: not the self or the thing (say, a cube) as already stabilized entities, but individuation, the formative (pre-individual) movement that also preoccupied artists like Klee: a "wild being" that envelops and incorporates, folding and unfolding topologically within the weave of Being. "Carnal being" thus has strata and depths.

Beyond any taste for literary figure, Merleau-Ponty, who maintains that "philosophy is not a lexicon," is not concerned with substituting the world by words: "it is the things themselves, from the depths of their silence, that it wishes to bring to expression" (VI, 4). Depth, abyss, profundity of Being: the things themselves require us to intertwine the epistemological level, an effort to go beyond language to the sensible, with the ontological: the deep, vertical, wild being. Hence the visible world is part of a process that encompasses both world and self within an indivisible relational structure articulated through multiple foldings. Moving beyond direct experience, however—the proper domain of phenomenology—means proceeding scientifically, by theory and experiment. To thematize the intertwining of epistemology and ontology calls for bringing philosophy into close contact with the sciences, aiming at an understanding of transphenomenal reality; yet a counter-movement is also possible, from scientifically investigated reality back to the world and, progressively, to subjectivity, returning mathematical and scientific speculation to the order of meaning, i.e., to phenomenological relations. Such unification of knowledge is, ultimately, philosophy's highest task.

What, then, is the relation between meaning, the world of experience, and the *punctum caecum*? Here the transcendental and the empirical intersect. Phenomenological constitution proceeds by conceptual clarification, bringing to light the implicit, making explicit the mediations of the immediate, rendering visible what does not directly appear. The subject cannot achieve full self-transparency: its contact with the primordial world is not transparent to itself; its inscription in reality via the world necessarily includes the blind spot. Hence the analyses of "nature" and "world" must illuminate the subject in its relational constitution: its lack of self-sufficiency is not a provisional limit of inquiry, but pertains to the essence of subjectivity, intelligible only through its constitutive bond with world and, ontologically, with reality.

To analyse the sensible is to descend into the primordial, pre-individual domain: from consciousness to the unconscious, to the deep logics of the sentient body: "the unconscious of consciousness (its central *punctum caecum*, that blindness that makes it consciousness i.e. an indirect and *inverted* grasp of all things) is the *other side* or the *reverse* (or the other dimensionality) of sensible being. (…) The flesh *is a mirror phenomenon*" (VI, 255). A Lacanian example illuminates Merleau-Ponty's terrain when, from "flesh," he articulates the gaze. In *Seminar XI*, Lacan analyses the gaze, Lacan analyses the gaze, explicitly citing his late friend. Consider a work which, at first glance, seems far removed from these concerns: Hans Holbein's *The Ambassadors* (1533). At first glance, a canonical portrait; yet at its centre harbours an anomalous object, an "object a."

The frame invites our entrance, guiding our gaze much as the physiology of the eye frames the visual field. But at the centre of our existence lies what we cannot (or must not) look at directly. The frame-eye opens onto and guides us through the world of life: here too, we see a world that ceaselessly sustains the meaning of our actions within the surrounding environment. If the frame-eye were to dilate to include its opposite pole, what then would come into view?

Stepping aside, the anamorphic blur resolves into a skull. Abandoning a frontal perspective, a "frontal ontology," for an oblique, "lateral ontology," yields a different image: the central enigma becomes death. Every worldly or scientific object, every vanity in the painting, loses force before this annihilating limit. Death cannot be lived or made visible, and yet it occupies the centre of our existence in symbolic guise. Here the eye (as frame) and gaze (as opening into the world) come apart. A purely frontal ontology cannot grasp the being-sense of what lies at the work's centre. We are not naturally oriented to negate the positivity of our existence; our comportment is oriented toward sense. Bringing Nietzsche into play, we may ask which "reactive forces" sustain the "conditions of preservation" of human life. If these preindividual forces orient sense and action independently of us, in what way can agency be free? Human being and its world are constituted in the reactive becoming of forces; directionality orients the gaze, activating drives (Freud) or desiring-machines (Deleuze) that operate upon the field in active–reactive modes, binding the subject to its milieu.

The world, as the other pole of subjectivity, forms with it an inseparable structure, our ecological niche. Merleau-Ponty's claim that we are "condemned to meaning" also bears on the world's character: we live within the frame-eye; all meaning is inscribed therein; beyond the frame lies the annihilation of subjectivity, consubstantial with existence. Death limits our life as perspective limits the appearing of the thing: such limitation is the very condition of knowing. Death, an invisible point of transcendence, marks and delimits the possibility of temporal existence. It stands at the centre of our field as the event in which life mirrors itself: there is death because there is becoming; we exist in continuous transformation and arrangement. What made birth possible and, symmetrically, its opposite, our finite span, must be oriented toward meaning. Hence the necessity of a *punctum caecum.*

For Merleau-Ponty, the point is not that everything is meaning or non-meaning but that, ontologically, there is meaning. There is no "world out there" except for us who live it in the first person: the "outside," reality, is the "inhuman" domain of scientific inquiry. To seek its meaning is to project onto reality predicates that belong to the world; in itself, it has none. We are "condemned to meaning" because we live in the world, in that niche of reality that emerges from the relation between our body and the surrounding physical environment.

A cut in the canvas, as in Fontana, does not traverse meaning or exceed observability. To slash the canvas—like the cut across the eyeball in Buñuel's *Un chien andalou* (1929)—does not, except metaphorically, surpass the visible horizon. Seeking direct access to what lies "beyond" the world is as senseless as attempting to leap over one's shadow. We may open ourselves to infinitely many digital worlds; we still encounter worlds, not a view from nowhere. Like the onion's layers, searching for a core is a misdirected metaphysics. The world is already reality; there is no "beyond" needed to find it. Precisely because, from the outset, we are part of it, relationally, we can question it. Were we *outside* truth rather than *within* it, we could never enter the house of Being nor partake in its manifold virtualities.

Chapter 11
The Dynamics of Figure and Ground

"Don't think, but look!"–Merleau-Ponty would probably have endorsed this motto from Wittgenstein's *Philosophical Investigations*. Having clarified the notion of the "perceptual field", we can now examine what unfolds within it at its most elementary level by analysing the figure–ground relation. First studied by Rubin in 1915 and then deepened by Koffka within the epistemology of Gestalt theory, this phenomenon yields a set of qualities and functional effects that distinguish figure from ground: unlike grounds, figures have object-character, exhibit epiphanic color, are more easily localisable in perceptual space, are semantically richer, and are therefore better remembered. Figures carry an attribute of "thingness" (weight and evidential force), whereas ground acquires the more "evanescent" features of the "non-thing".[1]

In pictorial representation, figures contrast with grounds along several dimensions: *relative size* (ceteris paribus, the smaller area tends to emerge as figure); *orientation* (regions whose axes align with the primary spatial directions, vertical and horizontal, tend to appear as figure); *inclusion* (the included region is seen as figure, the including region as ground); *convexity* (convex regions are preferentially figural); and *symmetry* (symmetric regions are seen as figure, while asymmetries favour a ground that appears to "pass behind," as in amodal completion).[2] These determinants must not tempt us to treat ground as a neutral, formless residue excluded from figural organisation. Ground functions as a *frame of reference.*[3] The figure–ground relation is the precondition for the phenomenal *unity* of the image. Any perceived object bears its own form and unity, which set it off from its context and confer position, orientation, and size.

Consider Rubin's vase–profiles (1921): contiguous, homogeneous black and white areas on the same plane. One perceives either a black vase on white ground or two white profiles on black ground. Contours operate unilaterally: either the vase or the profiles appear. By the principle of non-contradiction, the manifestation of one

---

[1] K. Koffka, *Principles of Gestalt Psychology*, pp. 185, 275 ff.

[2] Cf. G. Kanizsa, *Grammatica del vedere*, Bologna: Il Mulino, 1997, pp. 40 sgg.

[3] Cf. Metzger, *Psychologie*, p. 86.

L. Taddio, *The Appearing of Sense*, Lecture Notes in Morphogenesis,
https://doi.org/10.1007/978-3-032-27021-4_11

excludes the other; one cannot see "the same thing" as both identical and different at once. Although the distal stimulation is identical, we experience two distinct things: what is seen (one thing or two) and the meaning assigned to it differ entirely. As soon as one aspect becomes figure, the other becomes ground, and conversely. Crucially, the bistability is located solely on the phenomenal plane of direct experience; at the level of the proximal stimulus (retina), there is no bistability.

The articulation of figure and ground follows a precise phenomenological grammar: rules of self-determination, dynamic-relational processes internal to the phenomenal field. Things in the environment may appear relatively stable, or, where multistability obtains, ambiguous in ways that can be described as "objectively" ambiguous. Painters exploit this relation toward either stability or instability. Escher, for instance, constructs images that deliberately harness figural determinants to produce ambiguity first at the perceptual–expressive level and then at the level of meaning.

Take another case, cited by Merleau-Ponty: a white patch on a homogeneous background. Perceptually, every point of the patch belongs to the patch. Each point shares with the others a common feature by virtue of "a certain 'function' in common, that of forming themselves into a 'shape'" (PP, 4). The patch's color often appears "denser" than the background. Since the perimeter belongs to the patch, not to the ground, the edge is decisive in its definition. The figure's presence implies a background that continues behind it; we do not perceive the background as ending at the figure's border. "Each part arouses more the expectation of more than it contains, and this elementary perception is therefore already charged with a *meaning*" (PP, 4).

Here Merleau-Ponty isolates a decisive Gestalt insight: the relative autonomy of phenomenal-field functions *vis-à-vis* the conditions that occasion them. The task is not to posit formal a priori categories, but to identify the (transcendental) conditions which, within the field's organisational dynamics, render appearing possible. "Transcendental" is to be understood morphogenetically, as the genesis of meaning internal to the perceptual field. In this sense (with a modest hermeneutic stretch), we can read Merleau-Ponty's claim that "phenomenology, alone of all philosophies, talks about a transcendental field": we do not grasp the thing as an invariant object, but through the field's dynamic relations. "Phenomenology is a phenomenology, that is, a study of the advent of being to consciousness, instead of presuming its possibility" (PP, 71). We apprehend emergence through the field's self-organisation of forms (*Gestalten*): the invisible conditions the manifestation without being an objective datum; no distinct distal stimulus corresponds to the phenomenon.

It is not isolated things that determine the field's internal relations; rather, things arise from relational processes. Nothing appears without a background; appearing–existing entails a fundamental relation, and relation implies structure. Generative forces interact within the field, between things; these dynamic structures (as described by Gestalt theory) do not pre-exist as timeless models or archetypes but occur immanently, with and within the process. In this sense, we may speak of a "genesis" of the transcendental.

Return to the "Kanizsa triangle": one sees a figure, a triangle resting upon another. The upper triangle's virtuality belongs to the world no less than the lower one, though their modes of givenness differ. Such a Gestalt emerges from precise internal relations; it is anything but arbitrary. The Kanizsa configuration is as much part of our environment as a reflection, a shadow, a digital scene, or mountains and tables.

Ontologically, a thing's characteristics and expressive qualities obtain on the phenomenological plane—that is, in what is directly observed—rather than at the level of distal stimulation or retinal imaging. The field's autopoietic forms do not depend, strictly speaking, on will or decision. As Merleau-Ponty insists, both empiricism and intellectualism fail: each isolates a single pole (the "objective" *a posteriori* or the "subjective" *a priori*) and thereby overlooks the relational grammar operative in the field. The order of meaning in phenomenal appearing collapses when analysis fixes upon only one pole: it is the relational process that makes appearing possible. How, by analysing only parts, could one account for a background that "passes behind" the figure? The background terminates at the figure's contour and yet, phenomenally, continues beyond it: "Each part arouses more the expectation of more than it contains." Neither empiricism nor rationalism (in the broader sense of "intellectualism") captures this grammar. The field cannot be understood through abstract principles alone, without acknowledging its friction and resistance, its own organisational laws. The "more" that is seen (the background continuing behind the figure) is neither ontologically present in distal stimulation nor recoverable through the analysis of sensations alone. The phenomenon takes shape from relation as such, and relation must therefore be recognised as constitutive of appearing.

Gestalt psychologists, Merleau-Ponty observes, rightly maintain that figure–ground is not the "simplest" analysable case, a contingent example among others, but "the very definition of the phenomenon of perception, that without which a phenomenon cannot be said to be perception at all" (PP, 4). Only by apprehending field-organisation, the *Gestalt* internal to the perceptual field, can the sense of perception be grasped.[4]

The dynamic self-organisation of the field cannot be reduced to "sensation" (empiricism's basic unit) without compromising the very analysis of "perception" (PP, 8). Such reduction inevitably courts the *stimulus error*: sensation is a theoretical posit of science (retinal physiology, the physical object), not a deliverance of lived experience. The characteristics that emerge in figure–ground are not present in isolated sensations; sensation does not register the "more" beyond the stimulus that belongs to the phenomenon. Figures have "contours" that cannot be ascribed to ground; we see the phenomenon as "stable": the figure's color appears "compact", the ground "unbounded", its color "uncertain", even though we know it continues

[4] Merleau-Ponty therefore denies that a completely homogeneous area could be given to perception (PP, 4). A purported counterexample is the Ganzfeld or "fog effect" (Metzger), where the eye receives the same light in every direction. Under such sensory deprivation, higher cognitive functions (thinking, language, etc.) regress, indicating the cogito's only relative autonomy with respect to perception. We cannot know the world apart from the interplay of thought and perception; and, as Merleau-Ponty notes, "structure of actual perception alone can teach us what perception is" (*ibid.*). Philosophy cannot "teach" us the reality of things-in-themselves, which are not directly accessible; only science can, at least in part. Philosophy's task is to understand the sciences, precisely by recovering the "transcendental" dimension.

behind the figure. To apprehend meaning's articulation in the figure, we must take it as structure (*Gestalt*), that is, as figure–ground *relation.*

The thing, as correlate of experience, does not exist "in itself." Its appearing characteristics, those by which we identify and name it, exist only on the immanent plane of phenomenal experience. Even the claim that a thing is "isolated" is only ever relative. A figure without a ground cannot be given, not even in imagination. Grasping the thing's "meaning" requires acknowledging the complexity proper to direct experience, the very object of a phenomenology of perception.

Consider one final case. Merleau-Ponty recalls the apparent enlargement of the moon at the horizon compared to the zenith. To adopt empiricist or intellectualist presuppositions is akin to viewing through a telescope or cardboard tube: one loses the originally observed phenomenon, since the moon then appears at a constant size. To account for the phenomenon, one cannot alter the conditions of observability or replace the implicit frame of reference, our body and immediate perception. Empiricism "is not concerned with what we see, but with what we ought to see, according to the retinal image," while intellectualism misses appearing by "describes *de facto* perception according to the data of 'analytical' and attentive perception, in which the moon in fact resumes its *true* apparent diameter" (PP, 36).

Chapter 12
Visibility, Invisibility, and the Flesh of the World

Reality, taken ontologically, is a fabric that plunges beneath experience. In a working note from January 1960 in *The Visible and the Invisible*, Merleau-Ponty writes: "Principle: not to consider the invisible as an *other visible* 'possible,' or a 'possible' visible for an other; this would be to destroy the inner framework that joins us to it" (VI, 229). Our world and the worlds seen by others, he adds, are nonetheless "connected": they share a common fabric, a reality made of "inter-relations," which makes possible a world of dynamic relations internal to the field.

Art and perception draw upon the meaning of the world, not upon reality as such. If we apply a kind of phenomenological Occam's razor, this is what we may retain from Merleau-Ponty. The distinction between world and reality clarifies the interplay between science and philosophy (phenomenology). Following Merleau-Ponty's terminological suggestions, we may say that the visible–invisible relation can be thematised, on the one hand, as a relation of "ontological verticality" ("pure transcendence, without an ontic mask", VI, 229) opened up by scientific inquiry into reality, and, on the other hand, as a relation of "horizontality", through the analysis of that portion of reality we call "world," inseparably bound to our experience. The implicit frame of reference must be specified according to the question at stake: it is the problem that generates the method, not the method that dictates the problem. Philosophy and phenomenology cannot legitimately define the method once and for all and then apply it indiscriminately to "reality": the choice of conceptual tools depends on the order of problems to be addressed. To proceed otherwise would be like attempting to operate on a dental arch and on the heart with the same set of instruments: resorting to different conceptual devices does not undermine coherence; by contrast, the claim to apply a single method to problems of a different nature is what proves misguided.

Let us now consider the lifeworld. Merleau-Ponty notes that neither a painting nor a piece of music can be grasped in its idea except through the perceptual dimension and the articulation of colors and sounds. If we have not seen Cézanne's paintings, a purely conceptual analysis "leaves me with a choice between several possible

L. Taddio, *The Appearing of Sense*, Lecture Notes in Morphogenesis,
https://doi.org/10.1007/978-3-032-27021-4_12

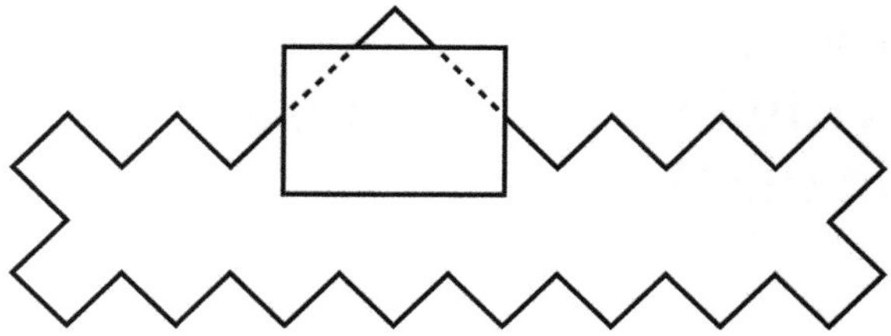

Fig. 12.1 Amodal perceptual completion

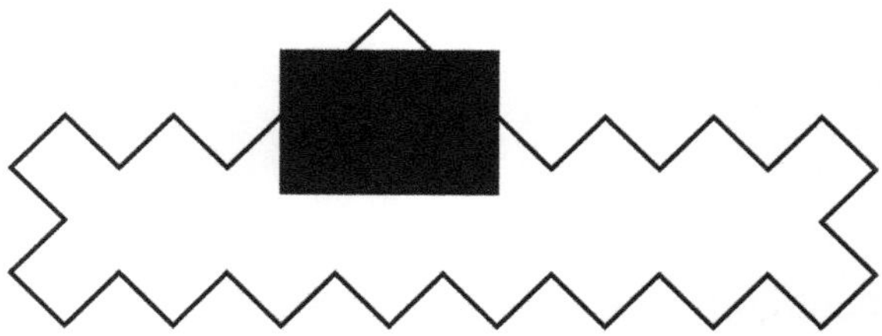

Fig. 12.2 Occluded configuration

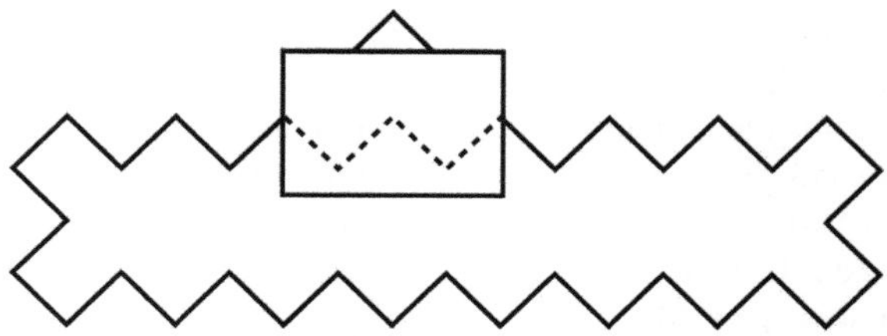

Fig. 12.3 Non-perceived logical completion

Cézannes": "it is the sight of the pictures which provides me with the only existing Cézanne, and therein the analyses find their full meaning. The same is true of a poem or a novel, although they are made up of words" (PP, 174). Art is nourished by the meaning given to us by the world, understood as the correlate of experience (in the sense distinguished above), not by reality as posited by the sciences. Consider now the following case of amodal completion, which allows us to give tangible content to Merleau-Ponty's claim that it is the perception of the painting (for example, *The Mont Sainte-Victoire*) that "which provides me with the only existing Cézanne" (Figs. 12.1, 12.2 and 12.3).

The reference figure is the central one: the task is to describe how we see the part hidden by the black tile. The two lateral figures offer two logically equivalent possibilities for completing the central configuration; indeed, the figure on the right might even appear "more logical," since the hypothetical completion, indicated by dashed lines, respects the structural coherence of the figure as a whole. Yet among all the logically possible completions, only one actually embodies the perceptual outcome (as shown in the figure on the left). Although many solutions are logically conceivable, only one is valid on the phenomenal–perceptual plane. Perception offers us *the* Cézanne that exists: within the phenomenal field there is a single way in which

Fig. 12.4 Ring-shaped figure with modal completion

we see the figure as amodally completed. The "invisible", partially occluded shape is not arbitrary. In principle, since it is not directly visible, we might imagine infinitely many completions, all equally coherent at a purely logical level. We *can* imagine various possible ways of connecting the non-visible segments. Nevertheless, we *see* only one solution, and this depends on the characteristics and relations of the directly visible parts: the visible part carries within itself the tendency to continue in a specific way and not in others, however theoretically possible. The choice "between several possible Cézannes" disappears when we stand before the painting, which offers us that determinate view, that determinate spectacle, and not any other. Just as in an electromagnetic field, within the phenomenal field invisible forces are at work which determine the organisation of forms. In his work at the canvas, the painter implicitly mobilises these forces of figural organisation; Gestalt psychology began to make them explicit, starting with Wertheimer's unifying factors.

Amodal completion thus shows that our "seeing the invisible" is paradoxical only from a logical standpoint, not from the standpoint of a phenomenology of perception. In everyday life we constantly encounter aspects of things that are not directly visible and yet belong constitutively to what is seen, as in the case of the cube or of the figure–ground relation. In all such cases there is no corresponding local stimulus for what is directly observed: the hidden part we see is not ontologically "there," just as the sides of the Kanizsa triangle are not "there," if we take the physical stimulus as our reference. And yet these aspects of the world are indeed there for us: we see them, we talk about them, and this experience can be shared intersubjectively. They belong to the phenomenal plane of experience and are part of the lifeworld. In this sense, the physical world and the phenomenal world are not superimposable: experience constitutes a plane of immanence from which we cannot leap into underlying or overarching reality (atoms or galaxies) except by way of indirect knowledge offered by the empirical sciences. This does not mean renouncing the role of observation within those sciences: we do, after all, observe experiments, and observation remains relevant even when the reality under investigation is not directly visible. Looking through a microscope, for instance, we do not *see* a micron directly, but an image several centimetres wide, which we *know* to correspond to a micron. Leaving aside the implications of observation in the natural sciences, let us return to completion. Consider Kanizsa's classic example (Figs. 12.4 and 12.5).

The three figures illustrate three different kinds of completion; only the third (Fig. 12.6) is a genuine case of amodal completion. In the first figure, on the left,

Fig. 12.5 Ring-shaped figure with mental completion

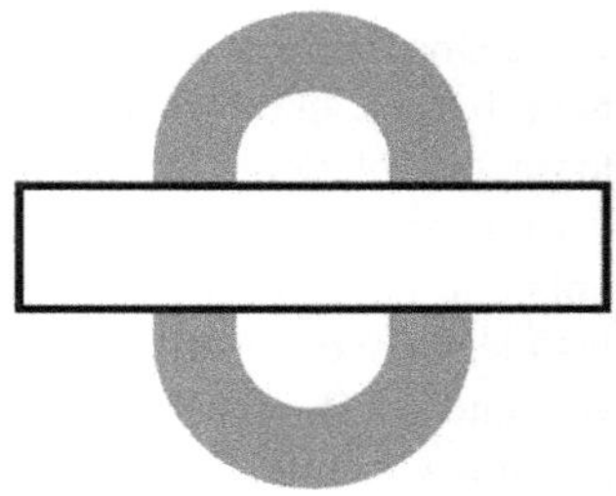

Fig. 12.6 Ring-shaped figure with amodal completion

the ring is completed modally; in the second, at the centre, we face a case of mental completion; in the third, the completion is perceptual. In the left-hand figure we see a single ring with no hidden parts, only junction marks. The unity of the central figure can be achieved by imagination or by cognitive integration; in both cases we still see *two* independent units, recognisable as distinct semicircles. The third configuration is different: here we do not perceive *two* separate shapes resting on a white rectangle, but *a single* figure that is partially hidden. We no longer have two independent units before our eyes, as in the previous case, but one occluded figure. We may imagine various possible ways of continuing the contours, all equally plausible in logical terms, but from a phenomenological standpoint only one completion is actually perceived. The nature of these three unities is therefore different: modal, cognitively integrated, and amodal. Amodal completion reveals the rules, constraints, and operative tendencies of the perceptual system.[1] Almost omnipresent in both paintings and everyday reality, it enables the direct perception of the invisible, regulating the relations of occlusion between percepts as they hide one another. It is called "amodal" because it has the character of *presence* (it is not merely supposed or imagined) even though it lacks the "normal" conditions of visibility (namely, a corresponding distal stimulus). The phenomenal *unity* of the figure is thus constituted by visible aspects that are *present* and visible aspects that are *absent*.

If we now turn to the surrounding world, the importance of amodal completion becomes evident: it is fundamental for the perceptual organisation of the unity of the thing and of the whole–part relation. Distinguishing things from one another, perceiving them as wholes in relative independence from their context, is for us

[1] G. Kanizsa, *Vedere e pensare*, Bologna: Il Mulino, 1991, pp. 51, 67, 31.

simple, spontaneous, and self-evident; it demands no explicit effort. Yet this ease is the result of a long process of evolutionary adaptation. Consider how complex it is, for example, to train a robot to grasp an object on a table in such a way that it recognises the object as an independent entity rather than as a segment of the table. That the thing on the table is a distinct and independent object is self-evident to us, but this self-evidence arises on the plane of perceptual organisation, on the phenomenological plane, prior to any act of conceptual classification. "Perception opens a window on to things", writes Merleau-Ponty: "it is directed, quasi-teleologically, towards a *truth in itself* in which the reason underlying all appearances is to be" (PP, 62).

Chapter 13
Illusion and the Structure of Appearance

"True philosophy consists in relearning to look at the world" (PP, xxiii). Merleau-Ponty teaches us to inhabit the ambiguities of body and perception; his oeuvre unfolds as an inexhaustible description of the phenomenal world. He is acutely aware of the "paradox" involved in bringing things into language, whose internal architecture cannot be clarified unless it is taken up in its complexity and followed through the unfolding of its articulations. This paradox, however, should not be read as an insurmountable obstacle to our access to the sensible-perceptual dimension: pain, for instance, needs no words in order to be felt. That language entails a dimension of interiority does not mean that it is sealed off or self-enclosed: "it is the subject's taking up of a position in the world of his meanings" (PP, 225).

By its very nature, language introduces a distance from the spectacle of the visible. Merleau-Ponty therefore invites us to recover the sensible thickness of things by taking as a model the painter's gaze upon the world; in other words, he shows that we can learn to see differently, to let the world come to expression through renewed perceptual attention. "There are many unclear sights, as for example a landscape on a misty day, but then we always say that no real landscape is in itself unclear. It is so only for us" (PP, 7). We must learn to apprehend the appearing of the thing in the manner proper to it, even when, as Metzger notes, the phenomenon strikes us as unusual, unexpected, illogical, or nonsensical, and even when it clashes with familiar conventions or entrenched habits of thought. We must learn to "let the things themselves speak", without allowing ourselves to be led astray by what is familiar, by what we have learned, by implicit knowledge, by the demands of logic, or by the poverty of our vocabulary.[1] It is not a question of dissolving perceptual ambiguities through logical tools, but of grasping the meaning that perception offers in the very way it appears: "This ambiguity is not some imperfection of consciousness or existence, but the definition of them" (PP, 387) (Fig. 13.1).

[1] Cf. Metzger, *Psychologie*, p. 15.

L. Taddio, *The Appearing of Sense*, Lecture Notes in Morphogenesis,
https://doi.org/10.1007/978-3-032-27021-4_13

Fig. 13.1 Zöllner's illusion

In the Zöllner illusion we see "the main lines converging" (PP, 40). "Intellectualism" attempts to reduce this illusory phenomenon to an "error" produced by the intervention of the auxiliary lines upon the principal ones. Yet what is required is precisely the opposite: to compare and analyse the principal lines in relation to one another. By "intellectualism", Merleau-Ponty refers to that long philosophical tradition, ranging from rationalism to critical philosophy, that conceives perception not *a posteriori* on the basis of experience, but *a priori* through the primacy of the *cogito*, judgment, or the intellect. Judgment is thereby assigned the task of dissolving the ambiguity of the figure. We are deceived, intellectualism claims, because we compare the entire set of lines (principal and auxiliary) rather than comparing only the parallel ones, falling thus into the "error" of perceiving a distortion relative to their "objective" parallelism. In the figure there are two spaces at play: "objective space," proper to geometry, in which the lines *are* parallel, and the *lived space* of direct perception, in which the same lines appear to converge. Why, then, is it so difficult to isolate (or "disincarnate") the principal lines from the auxiliary ones?

According to judgment, we should see them as parallel because we *know* them to be parallel. Yet, Merleau-Ponty observes: "Why the lines refuse to be separated from the auxiliary lines? It should be recognized that acquiring auxiliary lines, the main lines have ceased to be parallel" (PP, 41). Their perceptual "meaning" (that of being parallel) has been displace by another: that of divergence. The auxiliary lines do not merely superimpose themselves as discrete additions; rather, the meaning of the figure emerges *from their union*. Their insertion confers "a new meaning which henceforth clings to it and cannot be shifted" (PP, 41). It is the properties of this emergent phenomenon—the lines *as they now appear* to our gaze—that motivate the judgment we form concerning the "reality" of the parallel lines, in contrast to the supposed "appearance" of convergence. And yet, it is *precisely* from this observed divergence that one attempts to impose the status of "mere appearance", appealing to an underlying "objective" and "measurable" reality in which the lines are indeed parallel. The ruler placed upon the figure establishes geometrical objectivity, but the ruler cannot account for the ambiguity that *appears* to us.

If, instead of relying on judgment, we turn to empiricism, we might attempt to grasp the basic elements of the illusion by isolating them. Yet these isolated elements do not capture the phenomenon as it is apprehended in its unity. They become *something else* entirely. Phenomenology exceeds both empiricism and intellectualism. For empiricism, the thing is reconstructed from elementary sensations; the "thing" becomes the result of an assembly process rather than the primary, immediate datum. For intellectualism, the intellect determines perception. But then why does perception "refuse" to conform to the intellect's prescriptions? As our examples have shown, perception possesses its own rules of organisation, its own *autopoiesis*. For Merleau-Ponty, the task is to apprehend the appearing of the thing "over isolated impressions": this is what it means *to see*. The illusion consists in perceiving the "break up [of] the parallelism": intellectualism reconstructs the phenomenon through judgment, without truly understanding it (PP, 42). Judgment reduces the observed phenomenon to the categories of "true" and "false," but "must perception first be constituted" (PP, 42). Prior to the supposed underlying objective reality, which we affirm through the operationalization of scientific model, we encounter "a perceptual syntax constructed according to its own rules" (PP, 42). We can *grasp the meaning* that emerges from the *relations* present in the figure only on the perceptual plane, rather than attributing meaning from the outside (cf. PP, 43). The Zöllner illusion thus reopens "the problem of perception" (PP, 41), revealing the force of Metzger's injunction: we must learn to apprehend the thing as it appears, and in the very modes through which it appears, even if it appears "unusual, unexpected, illogical, or nonsensical."

Chapter 14
From Appearance to Judgment

Phenomenology offers the most compelling attempt to chart a path beyond both the "empiricist" and the "intellectualist" traditions. Our experience of things, contrary to what empiricism maintains, constantly urges us beyond their sensible manifestations; yet, against intellectualism, things are not exhausted by their appearances, nor are they mere units of order produced by judgment (cf. PP, 45–46). Still, the route opened by phenomenology is not without difficulties: the appearing of the thing is never fully resolved in the "here and now" perception of an aspect, nor solely through judgment. The form of the "thing" cannot be reduced either to consciousness or to an abstract synthesis of its constituent elements; it is through the "meaning bestowal" (*Sinngebung*) that such categories become possible.

Merleau-Ponty thus moves toward a phenomenological method that departs from analyses which disregard immediate experience—that sensible appearing from which all knowledge ultimately derives its meaning. There is no dichotomy between "aspect" and "thing", "appearance" and "reality", or "mind" and "body." Phenomenology sets aside the logical mediation characteristic of intellectualism while recovering naïve experience, apprehended in the subject–object relation and in the transcendence toward the thing. In naïve experience, things present themselves directly, mediated only by the mode of their own manifestation within a horizon of meaning. Our perspectival access to things reveals their "carnal reality": we know that what we see is only one possible aspect of a thing that transcends it. This transcendence is not *beyond* consciousness but *open* to it: this is the very definition of the "thing" for naïve consciousness (cf. SC, 237).

There is a certainty, a primitive "faith" in things that is shaken only afterward, once the *cogito* transforms this faith into an object of doubt. Likewise, the certainty of the external world falters when one attempts to ground knowledge in the "why" of things, through scientific explanation. Once things become objects of scientific representation, they lose the original source of meaning that arises from primordial contact with the world. Perceptual constancies provide stability across perspectival variations; reality, with respect to the world, does not consist in an underlying

L. Taddio, *The Appearing of Sense*, Lecture Notes in Morphogenesis,
https://doi.org/10.1007/978-3-032-27021-4_14

privileged substance but in a "framework of relations" (cf. PP, 349). Perception discloses this transcendence: it expresses, through language, the sensible knowledge that precedes the *cogito* and remains implicit until it becomes "figure," a thought capable of embodying the phenomenal life-world. For Merleau-Ponty, perception is the background "from which all acts stand out, and is presupposed by them" (PP, xi). The perceptual field, emerging from the relation between subject and environment, guides specific linguistic behaviours. Language thus constitutes an *open* system of expression (cf. PrP, 99, 109): though it refers to an interior dimension of the subject, it is always the other face of the world. Language expresses the stance of an embodied subject toward the world and its meanings (cf. PP, 207).

The perceptual "field" is not a mosaic of discrete sensations but a dynamic configuration conveying "functional values" integrated with forms. The whiteness of a sheet of paper in the shade, for instance, is not an objective datum but is perceived as white through an "all-embracing adherence to the world" (PP, 281). Merleau-Ponty aims to recover a pre-categorial knowledge that precedes word and concept, an opening to the truth of the *logos*: an implicit sensible knowledge, the direct expression of the empirical-phenomenal *continuum* in which the thing emerges between the visible and the invisible. The transcendence of the thing becomes manifest in the painter's activity, where expression, body, and space of the thing come to visibility.

The analysis of the Zöllner illusion has highlighted the insufficiency of both intellectualism and empiricism for *understanding* perception. Although the "real thing" and the "illusion" have different structures (PP, 395), "there could not possibly be error where there is not yet truth" (PP, 401). Illusions acquire their specificity within the order of the visible only through an evaluative stance adopted by the subject. Ontologically, the empirical-phenomenal plane is characterised not by the opposition appearance–reality, but by *virtuality*: whether a stone, its reflection, or its shadow, each appearance is a virtuality of Being. Only judgments, given their implicit frame of reference, can be true or false, never appearances as such. To subsume perception in judgment *presupposes* a sensible substratum with objective space (primary qualities) upon which a "second operation" *subsequently* distorts reality, generating the illusion via the auxiliary lines. This *approach* attempts to dismantle the illusion analytically, reducing the thing to its isolated parts. But this betrays the relational nature of perceptual organisation. The model of the living is not that of a "machine" whose functioning is revealed through disassembly: such reduction fails to grasp the nature of becoming and, consequently, the phenomenon itself. It would be like stripping an Escher drawing of its "contradictions": one would lose its very meaning.

Within the visual field, the principal and auxiliary lines of the Zöllner figure acquire an internal organisation wholly irreducible to the fragmentation imposed by judgment into a "real" and an "apparent" realm. There is no underlying reality beneath appearance. The illusion manifests its ambiguity in a visible, intersubjectively shareable way that no measurement can capture. To measure the illusion is to lose it: its phenomenal nature exists only in the relation established by our gaze. Instrumental measurement introduces a dichotomy between appearance and reality, reasserting the objective truth of parallel lines. But by proceeding in this manner, we are no longer analysing the same phenomenon. A phenomenological stance rejects this dualism: there are no "real lines" behind "apparent" ones, only the lines in their ambiguous obliqueness. "It is when we look away from them that they move insidiously towards this new relationship. There is here, preceding objective relations, a perceptual syntax constructed according to its own rules" (PP, 42). The dynamic rules of self-organisation in the perceptual field generate the very relations we perceive. The categories of "true" and "false," even when pragmatically invoked, do not constitute the illusion; they presuppose it: "in this way must perception first be constituted for predication to be possible" (PP, 42). As a perceptual phenomenon, the illusion *offers* a meaning graspable only on the plane of experience; judgment is merely "the optional expression" (PP, 40) of this meaning.

The Zöllner illusion thus reveals our tendency to evaluate phenomena through the implicit lens of "objective space," rather than receiving them in their appearing. More broadly, conditioned by this stance, we have lost the capacity to truly *see* the world. We relate to it primarily to transform and manipulate it according to our needs; scientific knowledge renders natural phenomena, in principle, infinitely manipulable. The digital domain exemplifies this manipulability: we can now inhabit worlds populated by "fantastic" objects, where the natural world becomes merely one among others. In this sense, as Nietzsche foresaw, the world has become a fable. To manipulate the world is to make it pliable, bending it to our desires and adaptive capacities.

Throughout human history, the natural environment has posed threats to survival. To transform it is to know it (*episteme*) and to control it. Scientific understanding entails the power to predict and anticipate possible dangers: *scientia potentia est*. This technical-manipulative capacity rests upon the presumed objectivity of primary qualities. With Descartes and Galileo, as Husserl argues in *The Crisis of European Sciences*, a progressive distancing from the world begins, transforming it into "representation." If the world becomes our representation, it can be subjected to our control. This trajectory, from Galileo through Kant, culminates in today's digital environments, where the "true" world returns to us primarily through the scientific image of reality: "The relation between perception and science is the same as that between appearance and reality" (C, I, §3; my translation). When naïve contact with the world dissolves "into image," do the colors, shadows, and reflections of a landscape become mere appearance? The objectivity of science eliminates the subjective gaze and with it our lived perspective. The scientific model, a "thought of the object in general," does not coincide with our embodied vision but nevertheless pretends to replace it.

It is only from Descartes onward that the existence of the world can be doubted, that is, only after understanding that the external world has been grounded in subjectivity, articulated through the *ego cogito* as the foundation of knowledge. We thus find ourselves at an absolute "here" vis-à-vis an absolute "there," an unbridgeable distance that constitutes, in Kant's terms, a "scandal of reason." Merleau-Ponty seeks to dissolve the dichotomy between *res extensa* and *res cogitans* by introducing the *lived body* (*Leib*) as the hinge linking world and consciousness. Cartesian doubt concerning the external world arises simultaneously with Galileo's mathematization of the thing, inaugurating a radical distancing from the aesthetic-sensible dimension (secondary qualities) in favour of total objectification. If the senses occasionally deceive us, Descartes argues in his *Meditations*, we cannot trust them at all. Yet should we not consider that the senses provide precisely the "frame of reference" and network of implicit relations that phenomenological analysis can render explicit?

Chapter 15
Illusion and Reality

Let us examine a concrete case of an optical–geometric illusion in order to understand the ontological and epistemological assumptions underlying the claim that the senses deceive us, namely, the relationship between appearance and reality. Why do we describe this figure as an illusion (Fig. 1)?

The answer seems obvious: we see one segment as longer than the other when, in fact, measurement of what is known as the "Müller-Lyer illusion" shows that, contrary to appearance (the presumed deception of the senses), the two segments are exactly the same length. The illusion, we are told, consists in the fact that within the same figure there are two objectively equal segments, although we perceive them as unequal. This incongruity does not arise at a single moment: when we observe the figure as a whole, we see two different lengths; when we measure it, our gaze is directed not at the figure as such but at the ruler and the data it yields. On the one hand, we have the "objective datum"; on the other, "appearance". On this basis, we classify the phenomenon as an "illusion" and take it as proof that, in some circumstances, the senses can mislead us. The very notion of illusion rests on the opposition between "experience" and "measurement." The equality of the two segments, as revealed by the ruler, is taken to represent "objective reality", while appearance is relegated to the realm of "subjective experience", in which the two segments have different lengths. This opposition presupposes that we measure the same thing we perceive: namely, the figure. But are we really measuring the same figure? Are we measuring the figure that we name the "Müller-Lyer illusion"?

If it is an illusion, it assumes this status precisely because measurement reveals the equality of the segments. The same phenomenon, when measured and when experienced, provides us with data that seem to contradict each other. We believe we are measuring the Müller-Lyer illusion and detecting a discrepancy between what appears and what the ruler tells us. We then classify the phenomenon as illusory on the basis of this "contradiction." We treat the measurement as objective and the immediate experience of the figure as subjective. We are inclined to regard the ruler's verdict as "truer," "more real," and "more objective" than what appears to

L. Taddio, *The Appearing of Sense*, Lecture Notes in Morphogenesis,
https://doi.org/10.1007/978-3-032-27021-4_15

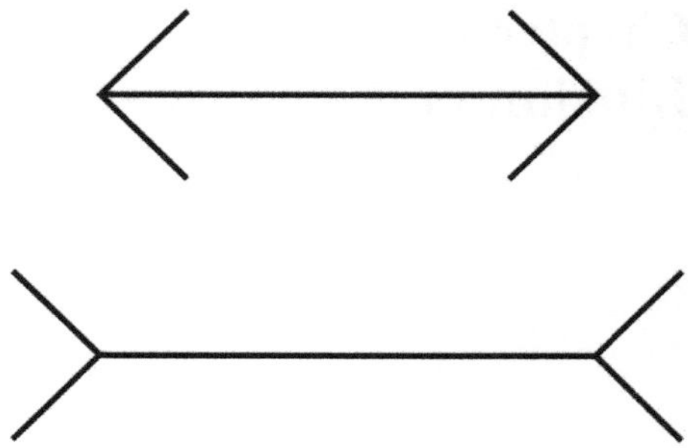

Fig. 1 Müller-Lyer illusion

us. Experience can deceive us; the ruler cannot. As a hypothesis, let us invert the relationship between appearance and reality. Could we regard appearance as "real", and the measurement as "subjective"? The measured data are obtained by observing not the figure, but the ruler. In this sense, we never step outside our condition as "observers": we remain within immediate experience of the world.

We consider one observed datum (that of the ruler) more real than another, but on what grounds? We might argue that the senses can deceive us, whereas measurement cannot. Yet we rely on the very same senses to see the measured data, and they could, in principle, deceive us here as well. What we should acknowledge is that it is not perception or appearance as such that deceives, but certain *judgments* we make about figures, images, or events. We interpret the figure as a trick played by the senses, whereas it is in fact a datum just as evident as that provided by the ruler. What do we actually do when we measure? We take the ruler, place it on the figure, read the length, take in the information from this act of measurement, move the ruler to the other segment, read the length again, extract a second piece of information, compare the two results, and finally formulate the judgment: since the segments are equal, the figure reveals a deception of the senses. The judgment is the final moment of a chain of operations; it is what establishes, on the basis of what has been observed, the illusory character of the phenomenon. It is based on the measuring operation carried out in two stages and on the comparison between those measured values and the *here-and-now* experience of the figure. In order to measure the phenomenon and to classify it as illusory, the Müller-Lyer figure, taken in its entirety and expressive richness, has been stripped down. Measurement entails an abstraction from the appearance of the thing. The ruler disregards the appendages that constitute the very *conditions of the illusion's appearance*, the "distortion" that affects the two segments. The ruler does not measure the phenomenon called the "Müller-Lyer illusion", but only the lengths of the two segments. Through measurement we abstractly isolate certain aspects of the figure, simplifying it for a specific purpose, namely obtaining the numeric length of the segments. Without the appendages, not only are the two segments equal in length, they also appear equal.

In a world made of painting, would appearance take the place of reality? Is the appearance–reality dichotomy reversible, depending on the frame of reference presupposed by our judgments? In the cases examined, the illusion is reality, or rather, it is part of that virtuality of reality that we call the world. In a drawn cube, what would be a "distortion" in terms of physical stimulus is precisely what allows us to see the object as proportionate. In such cases, as in the Müller-Lyer illusion, we experience the reality of the illusion *against* the ruler, against the measured datum.

To say that we "measure" the Müller-Lyer illusion is, strictly speaking, improper: we do not measure that figure which, taken as a whole, we have named the "Müller-Lyer illusion." The expression "illusion" serves to classify a set of phenomena that share certain characteristics. Yet the fact remains that it is impossible to "measure the illusion" with a ruler: the phenomenon, grasped in its full complexity, resists such reduction. Another kind of measurement is, however, possible, one obtained through direct experience: by systematically varying the position, shape, and inclination of the appendages, we can determine whether we continue to perceive one segment as longer than the other. This "measurement" is not obtained from the ruler but derives from the immediate experience of the event as it varies systematically, until we identify the configuration at which the "illusory" effect is maximal.

What we observe depends on the intrinsic characteristics of the figure, not on our "subjective disposition." If we alter the configuration of the appendages and the length of the segments, we modify the expressive effect of the illusion; if we vary our intentionality or mental set, the phenomenon does not change its appearance. In summary: (a) the illusion as such is "measurable" by the eyes but not by the ruler; (b) the relation between parts (appendages) and whole is a dynamic phenomenon. The analysis of the Müller-Lyer illusion allows us to highlight several epistemological features: (a) the event is repeatable; (b) the event is intersubjectively shareable; (c) even after measuring and *knowing* that the segments are equal, we continue to see one as longer and the other as shorter–which indicates a certain independence of the percept from the concept. Perception as such does not belong to the "truth value" established by judgment, which always presupposes a specific frame of reference. The Müller-Lyer phenomenon is real in that it is repeatable and intersubjectively shareable: these are two indicators of reality that scientific inquiry can hardly ignore, and they are the very criteria that define objectivity in the scientific domain.

"The two straight lines in Müller-Lyer illusion", Merleau-Ponty states, "are neither equal nor unequal length" (PP, 6). This *alternative* arises only if we continue to take the objective world as our reference frame. The remark is deliberately disorienting, for it clearly conflicts with our immediate experience of the figure; Merleau-Ponty's aim is to show that whatever judgment we make depends upon an implicit frame of reference. There exists a world and a "primitive perception" of it that precede all judgment.

The decisive question may be formulated as follows: are the two segments ontologically equal, or is the relationship between appearance and reality reversible? Does it depend on our epistemological stance, or is there, *within* the visible, an ontological substratum that determines the plane of reality in relation to appearance? Do the figures appear different while in reality being the same? Merleau-Ponty's theoretical intuition must be understood in these terms. There is, in principle, a form of "reversibility": the world, taken as appearance in its entirety, constitutes a plane of reality. To treat one part of appearance as "more real" depends on the stance we adopt toward appearance as such. Every virtuality of the real is real by definition; and one of these, the most significant for us, is what we call "the world." According to Merleau-Ponty, scientific knowledge of reality arises from "primitive" experience of the world. If the latter were incoherent with reality and not part of it, reality would remain inscrutable. If our subjectivity, by participating in the world, did not also participate in the real, the real would be not only unknowable, but inaccessible. Merleau-Ponty speaks of an "embodied" knowledge in which the *cogito* is not *separate* from reality, but the offspring of the same relations that constitute its fabric. In this sense, he affirms the "truth of perception" as access, participation, and inscription in the world in a way that is consistent with our form of life. If the world were other than truth, if experiences lacked coherence, both world and reality would remain beyond our reach.

To clarify the ontological status of illusions, let us consider the most famous monument of classical Greece: the Parthenon. Although based on precise numerical ratios, it was designed with perceptual considerations in mind. The mathematical and geometric regularities of the temple only fulfil their aesthetic function, in accordance with the classical canon of beauty, if they *appear* as such to the beholder. Vitruvius already knew that a building must take into account how it appears to those who see it if it is to be pleasing.[1] The body establishes the "perspective" and frame of reference through which we evaluate and appreciate it. Phidias and the other architects adopted this approach when they chose to "deform" certain elements (for example, the stylobate) in order to achieve an overall impression of regularity. These modifications aimed to correct "deceptions of vision" not on the basis of "objective" equality or *eurhythmy*, but on an equality and *eurhythmy* grounded in "subjective" experience. Ancient Greek architects introduced a series of corrections to compensate for "perspectival distortions". Their intention was, and still is, to offer the gaze "golden" canons: an appearance of the temple that is absolutely proportioned. Beauty is defined through "objective canons" proper to things that are in harmony with the *kosmos*. A work is "well made" when it is attuned to the whole. Yet, in order to produce this "objective beauty," Greek architects altered the objective proportions of the temple to make them consistent with how it is perceived. Through objective modifications grounded in the subjectivity of appearance, they conveyed what must appear as "objectively perfect". The concepts of "symmetry" and "proportion" at the basis of "objective beauty" thus paradoxically become real through illusion: the numerical ratios said to govern all things in art and nature are effectively realised in appearance.

The golden ratios that define the geometry of the Parthenon are accompanied by the deformation of the stylobate. The base of the temple is built with a convex centre so that it appears straight to the observer. Each column, in order to appear perfectly vertical, is constructed with a slight swelling at mid-shaft (*entasis*) and set on the stylobate with a slight inward inclination, so that it does not seem to open outward. The columns of the outer *peristyle* and the entablature are inclined

[1] Cf. M.P. Vitruvius, *Ten Books on Architecture*, trans. By I.D. Rowland, Cambridge: Cambridge University Press, 1999, Book IV.

inward to correct the impression of forward imbalance which, if the building were constructed "objectively" regular, would yield an optical effect of "irregularity." One illusion acts upon another, one degree of appearance of reality upon another. Something "objectively regular" that would appear irregular is deformed so that it appears regular, and this appearance is then elevated to an aesthetic *kanon*.

In the temple, some columns are built with a larger diameter (an objective deformation) in order for them to appear regular. Otherwise, lit in a certain way, they would seem thinner than the others. These optical corrections concern not only the architect but also the sculptor working on figures placed high above: the sculptor enlarges the head so that it appears proportionate to the observer on the ground.

For Merleau-Ponty, the "visual field" is a space in which apparently contradictory descriptions can coexist, because the elements in question–for example, the Müller-Lyer lines–"are not, in that field, assigned to the realm of being, in which a comparison would be possible, but each is taken in its private context as if it did not belong to the same universe as the other" (PP, 7). The object *in itself* "is never ambiguous," and the limits of the visual field do not fluctuate: an object approaching us, such as a distant train, suddenly appears at a certain point and then "begins to be seen." This is a well-known phenomenon in perceptual psychology: if we take an ink blot on a card and move it away from the observer, there is a distance at which it disappears and then reappears when brought closer. We do not perceive this broader threshold, and "attention" provides no testimony of the moment in which we become aware of it (cf. PP, 7).

Consider now the case of chromatic induction: a grey square of two square centimetres is placed at the centre of a blue square of fifteen square centimetres. We observe that the blue induces yellow in the grey square; if we reduce the size of the blue frame, the induced yellow becomes paler, so that we may say its intensity is a function of the blue. At a certain point, the blue frame becomes so small that the yellow disappears, leaving only grey. In this case, the apparent color is "too weak" to appear: the yellow ceases to be present and thus ceases to exist. To claim that the yellow still exists but is invisible because, as a function of the inducing area on the induced area, it now falls below a threshold of perceptibility is contradictory: the notion of a "sensation too small to be felt" violates the stimulus error. Inducing action exists only when one observes that something acts on something. The hypothesis that sensations below the threshold continue to diminish in accordance with the function established above the threshold has no phenomenological meaning: it is a contradiction in terms. To attest to the presence of yellow is equivalent to asserting an inducing action, but once the yellow is no longer observed, speaking of such an action is meaningless. By modifying the areas of the two squares, we retain a visible relationship between two states of affairs, while the supposed inducing influence becomes a representation, an object of thought that can be handled operationally and described by physics, but that no longer belongs to the phenomenological order. Within the frame of experience, perception is characterised by absolute independences: a phenomenon "begins" absolutely to be seen, in the sense that it either is or is not given. In his analysis of perception, Merleau-Ponty reveals the marginality of attention, defining it as an "auxiliary hypothesis" introduced to safeguard the "objective world": the phenomenon must be apprehended in its "indeterminacy"; its "meaning is an equivocal meaning; we are concerned with an expressive value rather than with logical signification" (PP, 7).

The analysis of illusions leads us back to Mach's thesis in *The Analysis of Sensations*, according to which the expression "deception of the senses" is imprecise. We have not yet fully recognised that the senses in themselves do not provide either correct or incorrect indications. "All that can be truly said of the sense-organs is that, *under different circumstances, they produce different sensations and perceptions.*"[2] When we say that we saw a man whereas "in reality" it was a scarecrow, we are making a judgment about an experience. It may seem "obvious" to say that the senses deceived us, since we believed we saw a man when in reality it was a scarecrow. But what has actually happened? In the meantime, we moved closer; the conditions under which the phenomenon appears have changed. We now see details that were previously invisible. We have not falsified the appearance as such, but our earlier judgment about it. It is not the percept that is "true" or "false", but the *judgment.*

The same holds for the classic case of the stick immersed in water, which appears bent when "in reality" it is straight. In the water, two distinct segments seem to appear, whereas "in reality" they are one and the same object. Ontologically, we might say that the stick is "one". Here too, we are judging two different situations: in the first case, the stick is in the water; in the second, it is outside. Its immersion provides us with a precise piece of information: *given* the laws of refraction, it appears as it must appear. If we assign an ontological status to every form of appearance, we find, on the same plane of immanence, both the broken and the whole stick, dreams and waking, the perceived surrounding world and the perceived in virtual environments. The appearance–reality dichotomy depends on the stance we adopt toward the world and its multiple manifestations. Yet not all aspects of the visible carry the same "weight" for us in terms of knowledge and adaptive behaviour: the visible is constantly subject to judgment and evaluation. Day by day, we see and judge the world. "Only the painter," says Merleau-Ponty, "is entitled to look at everything without being obliged to appraise what he sees" (EM, 161).

The two judgments about the stick, "whole" and "broken", refer to intrinsically different relations between thing and environment. Are we really comparing two judgments about what is, or what we consider to be, the same thing? Things constantly interact with their context. From a strictly perceptual point of view, it is not "the same thing," since the observed characteristics differ. If, however, we see the stick being placed in the water, we are disposed, given the short temporal interval and continuity of movement, to maintain that we are seeing the same stick go into the water.

Let us consider another example: we see an object so strongly lit or filtered that its color no longer matches the initial one. Ontologically, it becomes another object. We enter a room and see a green apple; we enter again and see it red. It might be the "same" object differently illuminated, or another, identically shaped but painted a

---

[2] E. Mach, *The Analysis of Sensations*, trans. by C.M. Williams, Chicago and London: Open Court, 1914, p. 10.

different color. The two situations, taken separately, do not allow us to decide whether it is the same object. In what sense, then, do we say we are before "the same thing"? Would it not, in all cases, be a different thing? As the studies on "naïve physics" show, classical physics is based on an intuitive and phenomenological structure: we tend to read reality in an intuitive and inductive way, extending to the real world the dimension of what is directly experienced (C, V, § 7).[3] We are all spontaneously Ptolemaic rather than Copernican: we see the sun move around us. To claim that one object remains the same while another has changed is to assign to the "material substrate", as in classical metaphysics, a privileged role, one that remains relative to the frame of reference adopted. From a phenomenological point of view, however, it is also true that, thanks to perceptual constancies (of shape, size, color, brightness, etc.), we usually see "the same thing" despite continuous environmental variations. When we see an object move, even as its perspective changes, we perceive it as the same object. A leaf falling in the wind is not, as in Borges's *Funes the Memorious* (1944), a different leaf at each instant, but the *same* leaf falling. To recognise that this feature of the world is tightly bound to our "image-apparatus" is not to deny that "something" would fall in the absence of a perceiver, but rather to deny that *this* leaf, with *these* colors and *this* expressive movement, would fall in the same way without any perceiving being. The question is self-contradictory, for the world itself, as the correlate of experience, would not exist without a subject who perceives it. Reality would exist, but not the world. To a naïve realist this may sound like a strongly anti-realist claim, but only if one fails to consider reality in the strict sense as a dynamic process of *relations*.

The leaf we see falling on a cinema screen is the result of a sequence of frames projected at a certain speed; below a given threshold, we would no longer perceive the continuity of the event, but only a series of still images. For us to adapt to that niche of reality we call "world," our experiential correlate must possess certain characteristics functional to our image-apparatus, just as the image-apparatus, emerging from a relationship and an adaptive history, is in turn functional to the world. There is a fundamental isomorphism rooted in the relationship that made possible the emergence of this world in relation to our "form of life." An insect would not see the falling leaf as we do, but the information it extracts and the information we extract are not in contradiction: different aspects of the same reality emerge in the animal world, to which we are reciprocally bound. As in the case of optical–geometric illusions, we face a situation that is intersubjectively shareable, repeatable, trans-historical, and trans-cultural. The phenomenon as such, what we see, precedes and accompanies our *interpretations*. We learn to look at and classify phenomena, not to see them. This originary character of appearing is veiled by the *use* of things understood as "means" for attaining certain ends through manipulation. To practise phenomenology is to relearn how to look at the world independently of the uses and evaluations tied to our power to transform it. It is to rediscover the world of perception, to observe and describe it without seeking anything beyond the spectacle that the sensible offers us, in the silent exercise that painters perform in their work. They bring into play the

[3] Cf. P. Bozzi, *Fisica ingenua*, Milano: Garzanti, 1990, Chap. 1.

intrinsic relations of the visible by letting the different elements of matter interact. Cézanne reveals things to us in the visible–invisible process of becoming: a multiplicity of equivalences of the world, a field of virtualities capable of preserving and generating new dynamics of meaning.

Chapter 16
Becoming and the Continuity of Experience

Time is not understood by Merleau-Ponty as a physiological fact, something reducible to or identifiable with specific parts of the body, nor as something centred on the Self. The perception of time is tied to the understanding of corporeality through an analysis of the temporal process within the perceptual field and its articulations. Time and movement are closely interconnected and must be understood within dynamic processes related to Gestalt factors and to what takes place in the "phenomenal present." Alongside time in the physical sense, there is therefore a phenomenological time, which must be analysed according to the attribution of meaning that characterises our existence, our actions, and our freedom to develop projects of existence within the world-environment. The appearing of movement is structured by a *logos* ("relation," "connection") that constitutes the condition for its being an "expression of movement".

Let us now consider some specific cases, beginning with the "stroboscopic movement." Here again we encounter the irreducibility of the *phenomenon* to the distal stimulus. We can observe stroboscopic movement when, for instance, we go to an amusement park and watch the apparent "flow" of small lights, or in the basic principle underlying cinema. In the era of film reels, during the projection of a movie we saw the flow of events in motion rather than individual static images. Once more, a difference emerges between the distal stimulus (the individual frames) and movement, that is, what occurs in the "phenomenal present" of the cinematic narrative. Similarly, in stroboscopic movement, if we analyse the phase of physical stimulation, we find two isolated lights that switch on and off independently at two different points in space. Yet, given an optimal *interstimulus interval* (ISI) of roughly 50–70 ms, we no longer see two independent light sources, each with its own phenomenal unity and identity, but *one* luminous ball moving from one point to another in space. The different situations can be described with respect to the first light (L1) and the second light (L2) as follows:

(1) We observe a movement from left to right.
(2) Yet lights L1 and L2 do not move.

L. Taddio, *The Appearing of Sense*, Lecture Notes in Morphogenesis,
https://doi.org/10.1007/978-3-032-27021-4_16

(3) First L1 turns on and then L2 turns on.
(4) I therefore call the switching on of L1 the *cause* of the movement and the occurrence of the movement the *effect*, because if L1 does not turn on, I see only L2 standing still in its place.
(5) But if L2 does not turn on, L1 also remains still in its place.
(6) I then call the switching on of L2 the *cause* of the movement.
(7) However, L2 turns on when L1 is already off, so it cannot have *caused* the movement of something that no longer exists.

In summary: what kind of "time" is this, in which what happens *afterwards* influences what happens *before*? Since movement has a duration, what kind of duration is it that begins with what happened later and ends with what happened earlier? The situation is paradoxical because, at first sight, everything appears as if L1 began to move *before* L2 was switched on, without "knowing" whether L2 would in fact turn on or not.[1] To recapitulate: there are two physically independent events (L1 and L2); both stimuli, taken individually, generate phenomena that do not imply any movement. At one location we see an event–the switching on and off of a light source–and we see the same thing occur at another location in perceptual space. These events are placed in a specific relation, a relation that yields certain numerical values. Those values, and no others, represent precise constraints: from this *relation* emerges the movement from L1 to L2, so that we are no longer dealing with two stationary events, but with a single event, a single "thing" moving from left to right. This event is part of the "world"; it exists within the world and arises under precise conditions: it emerges from a relation, from that interval of time between 50 and 70 ms which determines the unfolding of the appearance of movement. As in the organisation of figural units within the dynamics of the field, here too the formation of movement takes place on the basis of a specific numerical relation that structures (*Gestalt*) an event within dynamic aspects that determine the appearance of "apparent movement."

Merleau-Ponty discerns the same relational dynamic in the retroactive "relation" at work in melody, a classic example of Gestalt theory since von Ehrenfels: a melody is a *Gestalt* because it possesses a unity that cannot be reduced to the mere sum of its parts taken in isolation. Such isolation, in fact, prevents us from grasping the dynamics relations at play. Merleau-Ponty emphasises an additional aspect, and it is significant that he does so in his Collège de France course on *Nature* (1956–1960), where he analyses the concepts of movement, behaviour, and environment. Central here are his reflections on von Uexküll's concept of *Umwelt*, together with constant references to the work of Goldstein, among others. Stroboscopic movement is a phenomenon that Merleau-Ponty considers repeatedly, beginning with the *Phenomenology of Perception* and then, in a more systematic way, in his 1952–1953 course *The Sensible World and the World of Expression*. In relation to behaviour in the environment, as in perception, there are phenomena of feedback and reversibility that lead to a new conception of the body as a perceptual system that is both active and passive with respect to its milieu: information does not travel in a single direction

[1] Cf. G.B. Vicario, *Il tempo*, Bologna: Il Mulino, 2005, p. 187.

according to a rigid stimulus–response or cause–effect model. The world relates to our body through a continuous exchange that orients and determines it as a "perceptual system" in multisensory terms. To perceive the surrounding environment thus implies an osmotic exchange of information.

Drawing on Gibson's reflections, we can further enrich the framework outlined by Merleau-Ponty. The relation between light and a given layout specifies and determines a structure that carries invariant information for the perceiving organism. The optical array (and the optical flow), as defined by Gibson, is "structured," cannot be separated from the temporal dimension, and must therefore be understood as a perception–action *relation*: it contains "information" (derived from light) about certain characteristics of the environment within which we act and which we directly perceive in the world. Each of its expressive forms (affordances) in turn conditions our action in the environment, as if we were on a Möbius strip in which inside and outside are interwoven in a continuous movement. From this perspective, world and body are, in every respect, a dynamic form (*Gestalt*) that implies, as Merleau-Ponty says, a "chiasm," an "intertwining." In the same way, the beginning and end of the melody, as in stroboscopic movement, mutually imply one another.

The world-environment is composed of relatively stable phenomena, from the perception of the sea to the background murmur in a room. Yet if we focus our attention on certain aspects, we begin to see the waves and perceive their internal movement, just as, in the case of the murmur, we can cease to hear a "constant disturbance" and instead discern a sound phenomenon composed of recognisable elements such as voices and noises. If we analyse a musical scale, the stability of each individual note (assuming its frequency remains constant) dissolves within the form of the musical phenomenon perceived as ascending or descending. Thus, just as the sea is perceived as still while being composed of moving elements, we can perceive movement from elements that, taken individually, are static.[2]

In the same way that the ancients conceived individual parts in harmonic relation to the whole, Merleau-Ponty metaphorically conceives melody as the proper relation within the world-environment, itself governed by dynamic principles internal to the phenomenal and behavioural field. These dynamics are not unidirectional; they do not proceed merely from the physical stimulus to the body, which would then process it passively. Rather, the constituents reciprocally condition one another (reversibility): the relation between parts determines the emergent properties of the whole, and the perception of the whole retroactively transforms the function of the individual parts. In *The Visible and the Invisible*, Merleau-Ponty uses the image of a "two-leaf being" to evoke this circularity in the body–environment relation.

Since our body is a multisensory and sensorimotor[3] system integrated with the environment, phenomena feed back into and structure our behaviour through a continuous internal–external exchange of structured information (between activity and passivity) that maintains a certain degree of coherence: the notion of "Sign-Gestalt"

[2] Cf. G.B. Vicario, *Il tempo,* pp. 17, 19.

[3] Cf. N. Bruno, F. Pavani, *Perception: A Multisensory Perspective*, Oxford: Oxford University Press, 2018, Chap. 4.

captures precisely this retroactivity. By modifying, for example, the ending of a melody, we alter its overall aspect; something similar occurs within the dynamic structures underlying biological environments such as organisms (SB, 107). Such relations, in stroboscopic movement, in the retroactive effect present in melody, and in other experiments in Gestalt psychology (for instance, the "tunnel", "window", and "stop" effects),[4] lead us to understand the relation between time and movement as *revealing the being* we call "virtuality". One such virtuality, the "world" (the correlate of our experience) arises from relations internal to reality. The independence of the world and its relative stability (perception–change) are established by a certain metastable form of life: this world is the expression of our corporeality (*Gestalt*), understood as a perceptual system, and it manifests itself in our phenomenal present.

[4] Cf. G.B. Vicario, *Il tempo*, Chap. 5.

Chapter 17
Time, Movement, and Perceptual Genesis

The most vivid expression Merleau-Ponty uses to indicate the thrust of his attempt to shape a new ontology, drawn from the notes of his 1953 lectures, is that of movement as a "revelator of being" (*révélateur de l'être*). To understand this claim, we must turn in particular to lectures 7 and 8 of the *Course*, where the theme of "apparent movement" is addressed through the analysis of "stroboscopic movement." As the explicit references show, Merleau-Ponty's main source in these passages is Chap. 7 of Kurt Koffka's *The Principles of Psychology*. Several authors to whom Merleau-Ponty refers (for instance Wertheimer, Duncker, Ternus, and Metzger) are discussed there as well. What concerns us, however, is his reading of these experiments and of the related arguments, in order to show how Merleau-Ponty's phenomenological method—his way of working in close connection with experimental psychology—still represents today one of the most relevant models of phenomenology.

Starting from these Experiments, the Problem of the Ontological Status of "Apparent Movement" Emerges, and with it the Question of the Connection between Movement and Reality. Once Again, Merleau-Ponty Identifies the Key to this Problem in the Epistemological Implications Derived from the Analysis of the Experiments. Let us Consider some of them in Order to Clarify the Issue and Thereby Understand the Meaning of the Notion of *révélateur de L'être* Just Mentioned. We Will Limit ourselves to a Few Cases of Apparent Movement. The Central Point of this Reflection Lies in Grasping the Ontological Status and the Very Meaning of the Concept of "Appearance" in Relation to "Reality" in Movement. The Chiasm that Binds these Two Concepts Is Epistemologically *Reversible*. Whenever we Describe Part of the Visible as Merely "Apparent," this Gesture Necessarily Implies the Positing of a Plane of Reality in Relation to Appearance. The Task, Then, Is to Identify, through Merleau-Ponty's Own Examples and Observations, the Theoretical Cores that Allow us to Define the Terms "Appearance" and "Reality" and their Underlying Presuppositions.

As in the case of stroboscopic movement and its numerous experimental Analyses, the quality of the phenomenal aspect (our direct experience of the world) has a certain

L. Taddio, *The Appearing of Sense*, Lecture Notes in Morphogenesis,
https://doi.org/10.1007/978-3-032-27021-4_17

degree of independence from the underlying physical world. On this basis, Merleau-Ponty can state that "all movement is stroboscopic" (Sw, 58): apparent movement is a real movement understood as a "limit case" of "stroboscopic movement" (see SW, 150). Neuroscience has shown that a lesion in area V5 leads to the loss of the ability to see movement. The capacity to perceive movement may thus be understood as the result of a long evolutionary process that enables us to orient ourselves within the surrounding environment. Our basic perceptual and cognitive system is not primarily aimed at knowing "reality," but is the outcome of processes of adaptation.

We can implicitly draw an analogy between what happens on our retina and what occurs in stroboscopic movement. As previously noted, cinema offers a clear way of making this analogy intelligible by relating the cinematic experience to the individual frames: as is well known, stroboscopic movement lies at the basis of film projection. The fusion point is reached at around 24 frames per second, a temporal interval that produces the illusion of movement. What emerges—the cinematic continuum—possesses a degree of reality independent of the physical support that makes it possible, the film strip. For example, a scene that elicits fear is characterised by a specific event in the narrative of the film, and from an ontological point of view this narrative event is the true correlate of our feeling. The narrative present cannot be reduced to an underlying level: movement "unfolds" being, that is, it emerges from a plane of reality that has a certain independence from the physical level that conditions it.

For Merleau-Ponty, this going beyond the distal stimulus in the case of stroboscopic movement is comparable to the examples considered above involving figure and ground: when we see the background continuing underneath the figure, or when we consider the visual field as a space bounded by clear limits even though it extends behind us. In continuous movement, which we take as part of the phenomenal world, we perceive the unity of a movement that appears as such even in the absence of a corresponding movement at the retinal level (the distal stimulus). Cinema, in this sense, offers an effect of reality grounded in stroboscopic movement (see SW, 58). The point is not to identify some underlying reality, "true" reality in opposition to an apparent world; rather, the primary frame of reference is the "reality" of appearance itself (the world) from which we then proceed to investigate underlying physical reality.

In the domain of apparent movement, experimental measurements show that the distance between the lights (distal stimulus) does not affect reaction times, which are nonetheless longer at the moment the lights are switched off than at the moment they are switched on. Moreover, the greater the distance between the stimuli, the more the apparent speed of movement decreases. These *variables* do not depend on our subjective attitude. They are properties of the appearing of things and of their constitutive relations, which under certain conditions produce effects of reality that characterise the world and, under others, the "cinema".[1]

---

[1] Cf. G.B. Vicario, *Il tempo*, p. 186.

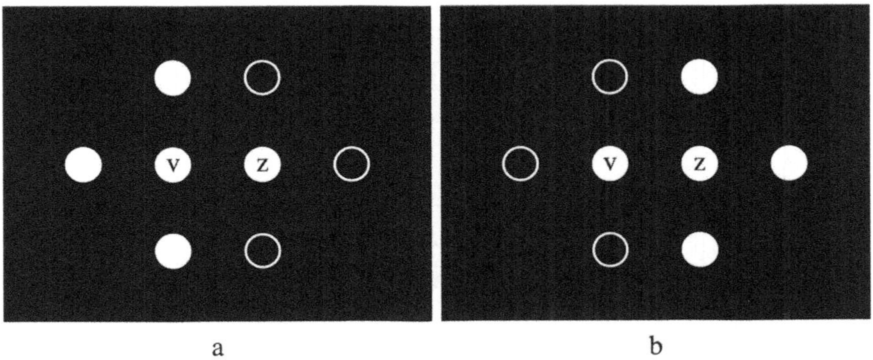

Fig. 17.1 Ternus experiment

Equally relevant in relation to stroboscopic movement are Ternus's experiments, which aim to highlight the Gestalt coherence of structure within the dynamics of movement.

In 1926, Ternus conducted a series of experiments on the preservation of the unity and identity of stimuli subjected to stroboscopic changes. The experimental setup involved the simultaneous presentation of multiple points that were switched on and off. For example, if the illumination of the five white points in Fig. 17.1a was followed, after a brief interval, by the illumination of the five white points in Fig. 17.1b, the entire configuration of five points was seen to shift from left to right, together with the two central points V and Z, which were lit in both phases. In fact, points V and Z underwent no change. Nevertheless, their displacement was perceived: perceptual activity tends, as far as possible, to preserve the unity and identity of forms.[2]

When the luminous points are arranged as in Fig. 17.2a in the first phase and as in Fig. 17.2b in the second, the five central points that are illuminated in both presentations should, from a purely physical point of view, appear stationary. In fact, one ought to see only the three leftmost points "jumping" to the right. Instead, what is perceived is that the five central points move along a curvilinear trajectory, until the upper-right point in Fig. 17.2a comes to occupy the lower-right position in Fig. 17.2b. This phenomenon highlights the extent to which perceptual activity strives to preserve a relatively stable environment through continuous transformations and displacements. For Merleau-Ponty, two consequences follow from this phenomenon: (1) The perceived movement of the arc is not given by "the sum of the movements of its parts;" it is rather the figure (*Gestalt*) as a whole that moves. The structure is not additive: "The movement flows from the whole arc (or cross) as such." We do not perceive an aggregate of displacements, but "the very apprehension of the figure," "the becoming of a figure." (2) The structure (*Gestalt*) does not first appear as a stable form and only subsequently undergo a displacement. Rather, we immediately

[2] Cf. G. Massironi, *Fenomenologia della percezione visiva*, Bologna: Il Mulino, 1998, Chap. 5.

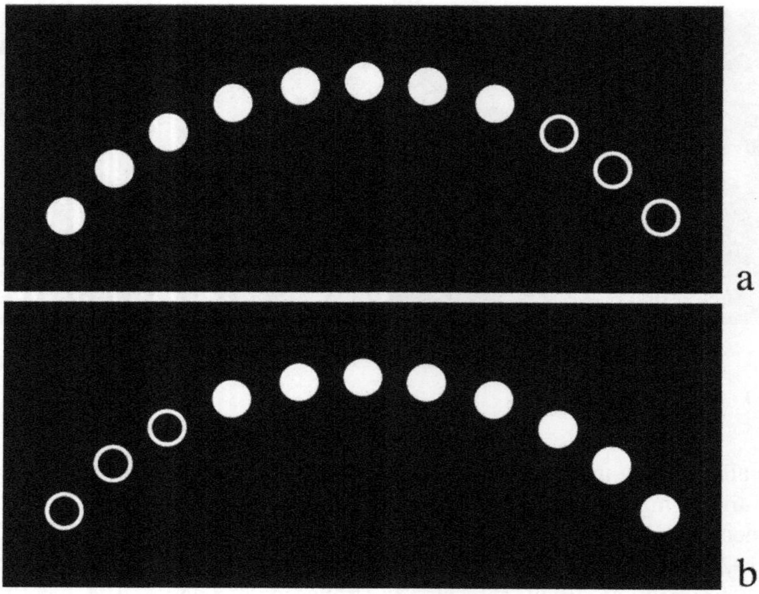

Fig. 17.2 Ternus experiment

see it passing from its initial position to its final one. Its identity is recognisable only through this becoming: the figure is the "trace of a certain type of movement, of certain dynamic properties" (SW, 60).

These examples aim to emphasise the decisive role of appearance within the dynamics of movement. Yet it is precisely the ontological status of what we call "appearance" that becomes problematic. Appearance is typically conceived in opposition to an underlying reality. From an ontological point of view, however, even the apparent stability of phenomena belongs to the very elements through which reality is identified as such. We must therefore acknowledge a divergence between what phenomenology designates as "reality" and what it designates as "appearance," a divergence that marks the relative autonomy of the phenomenal world and prevents it from being reduced to the physical plane.

Let us consider another case, that of "induced motion" analysed by Duncker in 1929. This phenomenon arises when the movement of one element in the perceptual field is attributed to another. In complete darkness, only a rectangle and a point inside it are visible (Fig. 17.3). When the rectangle is displaced to the right, the point, which in fact remains stationary, is perceived as sliding to the left, while the rectangle itself appears motionless. This occurs because perception spontaneously organises the scene by *differentiating systems* and establishing hierarchies within it. What appears to move is determined by the relation each element bears to the implicit frame of reference. Since this reference frame is tacitly taken to be more stable than

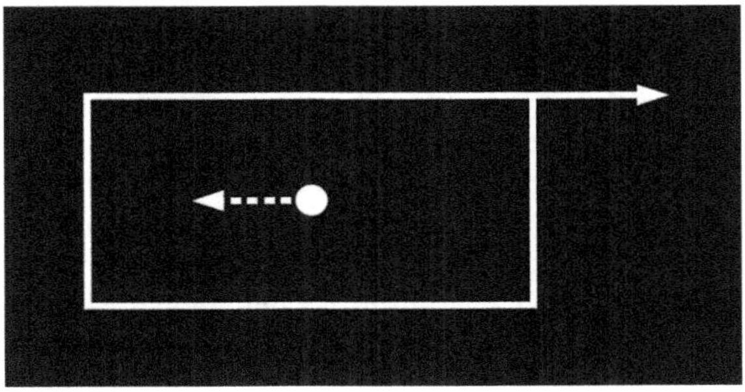

Fig. 17.3 Duncker: induced motion

the elements it contains, movement is attributed to the latter, unless information is available that disrupts this organisation of the field.

Merleau-Ponty does not merely acknowledge the importance of induced movement for a theory of perception; he radicalises it through a phenomenological description of the Gestalt as it is directly given in experience. "[E]ven if personal–historical factors are involved, they are only given to us as sedimented in the perceptual landscape" (SW, 64). Perception does not process data; it does not calculate. It immediately grasps a result. Nor is perception a "spiritual act": "organization is not [a] *Sinngebung* by us as [a] thinking subject but as [an] incarnate total being endowed with a certain past" (SW, 65). Perception is the meaningful configuration of the process as a whole. Scientific analyses may clarify the role of partial components, but they cannot account for perception as the immediate *Gestalt* of the world, our primordial mode of being in contact with it.

For this reason, "the apprehension of meaning and the apprehension of movement are the same thing". Perception allows us to "see meaning rather than seeing on the basis of meaning" (SW, 65); it must therefore not be conceived as the act of a detached subject. The genesis of movement is internal to the organisation of the field itself. Merleau-Ponty finds further confirmation of this view in Michotte's (1954) experiments on the perception of causality, especially in the so-called "launching effect". In this classic experiment, two rectangles, A (black) and B (red), are presented on a screen at a short distance from one another. A moves toward B at a constant speed and stops upon contact with it. Immediately afterward, B begins to move away in the same direction. Although we know that there is no physical transmission of force between the two shapes, we spontaneously describe the event as a causal sequence: the black rectangle hits the red one and pushes it away. What is perceived is not a mere succession of displacements, but a single dynamic event articulated in terms of impact, impulse, and transmission. Causality, in this case, is not extracted from the physical data but belongs to the expressive structure of the phenomenon itself.

In the same 1953 course, Merleau-Ponty explicitly establishes an equivalence between movement and expression: "movement = expression. (...) Hence [the] indefinite expressive capacity of movement. Here expression [is] defined as [the] emergence [apparition] of an existence" (SW, 150–151). Movement is not an external property added to being, but the very mode in which being manifests itself. In

Michotte's experiment, for instance, causality is not inferred as a conceptual relation; it is directly given as an expressive quality of the phenomenon (impact, pressure, propulsion). These expressive properties are not secondary or subjective overlays; they belong to the ontological texture of what appears. In this sense, movement has its own ontological dignity: it is a "revelator of being". To say that movement reveals being means that being is disclosed through expression. In other words, expressive qualities (affordances) stand on the same plane as so-called primary and secondary qualities. Appearance itself has ontological weight. In line with Merleau-Ponty's later ontology of the "wild" or "vertical" being (VI, 178), movement is not merely something that happens *to* beings; it is one of the foundational modes through which being comes to presence.

The phenomenal domain explored by Merleau-Ponty through these experimental references thus emerges as an authentic phenomenology developed in dialogue with the psychology laboratory. It is an experimental phenomenology capable of drawing out the ontological and theoretical implications of empirical findings. This method retains its full relevance: the experiment becomes a privileged site for philosophical interrogation. Phenomenology, in this sense, does not stand outside science, but operates from within it in order to bring to light the layers of meaning that scientific analysis alone cannot exhaust. In this course in particular, the outlines of a new ontology clearly take shape, an ontology in which movement, expression, and appearance are no longer relegated to the domain of the merely subjective or derivative, but are recognised as fundamental dimensions of being itself. Merleau-Ponty's analyses do not dissolve into experimental data; rather, they show that experimentation itself presupposes a field of meaning that it does not fully master. Immediate experience thus retains a certain autonomy, which in the final phase of his thought is radicalised into an explicitly ontological project: an ontology capable of overcoming the classical opposition between appearance and reality precisely through the notion of movement as *révélateur de l'être*.

Chapter 18
The Present as Duration

For a clearer grasp of stroboscopic movement, melody, and, more broadly, of time as a *structure* (*Gestalt*), let us consider the following diagram, which Merleau-Ponty borrows from Husserl.

The diagram shows that time cannot be conceived as a line composed of discrete events A, B, C. It depicts a horizontal line punctuated by successive "nows" (phenomenal presents). A is one such "now," in which something takes place "here and now"; B is another, and so on. The oblique lines represent the *retentions* of the same now as grasped by the following now, whereas the vertical lines represent the subsequent retentions of that same now (PP, 484). Applied to stroboscopic movement, this means: at first glance, the line appears fixed in position A and then, suddenly, appears in position B. But if we vary the temporal cadence, the stroboscopic movement disappears and we instead see two lines, either simultaneous or successive (PP, 313–314) (Figs. 18.1 and 18.2).

We are not confronted with the "passage of a line" but with a "pure passage." The example is instructive only if it is not interpreted as an "illusion" of movement but as the structural form of real movement in the lifeworld: for instance, when observing workers passing heavy goods from hand to hand, we often perceive only the initial and final positions of their arms, yet we apprehend the movement as a complete, unified event. Movement is not a chain of discrete instants, "an indefinite series of positions," but a phenomenon that presents itself in terms of beginning, continuation, and end (PP, 314). The same logic applies to melody. A melody is not the mere succession of notes but a *Gestalt*: an expressive form that is not *reversible.* A melody played backwards becomes *something else* entirely. Its directionality, its "temporal arrow," is intrinsic to its sense; it possesses a perceptual pregnance, as a *figure* on a ground, and must be apprehended as a whole: "each note only counts by virtue of the function it serves in the whole" (SNS, 49).

There is therefore no discrete series of isolated "nows" that one first registers as images and then subsequently links into a line. Each phenomenal instant has its own "thickness", and each is modified by what follows: the present retains the past and

L. Taddio, *The Appearing of Sense*, Lecture Notes in Morphogenesis,
https://doi.org/10.1007/978-3-032-27021-4_18

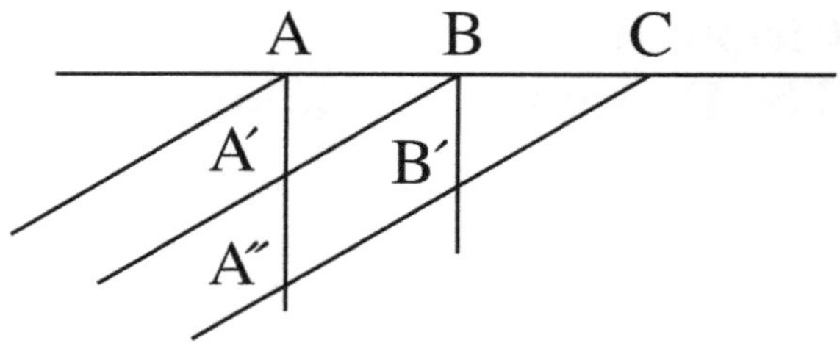

Fig. 18.1 Husserl's time-diagram

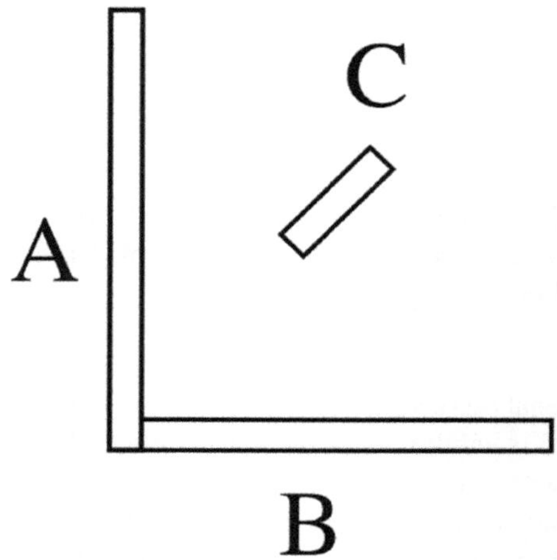

Fig. 18.2 Wertheimer: stroboscopic movement

anticipates the future. The present is not a punctual, empty moment; it is a *Gestalt*, a complex event composed of past and future (PP, 489). We do not perceive A, A', A" as isolated moments; rather, just as we can see a stone immersed in water through its distortions, we perceive A through A', and A' through A" and so on (PP, 485). As each moment arrives, it transforms the preceding one, deepening the temporal thickness between them. Husserl's diagram must therefore be completed symmetrically by protentions: "time is not a line, but a network of intentionalities" (PP, 484).

Gibson, in *The Ecological Approach to Visual Perception* ("The False Dichotomy Between Present and Past Experience"), isolates the same phenomenon that Merleau-Ponty emphasises. In the environment, no static boundary separates the perception of the "now" from that of the "just-past": the experience of time is not the recession of a past from an instantaneous present. If such a boundary existed, we should be able to say exactly when perception ends and memory begins.[1] But this line is nowhere to be found: duration is phenomenologically given even before any attempt to measure

[1] Cf. J.J. Gibson, *The Ecological Approach to Visual Perception* [1979], New York and London: Psychology Press, 2015, pp. 242–243.

it.[2] The replacement of memory by perception occurs within the same perceptual flow. Past, present, and future are abstractions: "what there really is, is not a past, present, and future." The present does not sit atop an accumulating past nor propel an awaiting future; rather, it is the continuous transition of future into present and present into past (PP, 486).

"The solution of all problems of transcendence", writes Merleau-Ponty, "is to be sought in the thickness of the pre-objective present, in which we find our bodily being, our social being, and the pre-existence of the world, that is, the starting point of 'explanations', in so far as they are legitimate–and at the same time the basis of our freedom" (PP, 503). The phenomenal present is not a mathematical instant, devoid of temporal extension, but an articulated event composed of retention (past) and protention (future). The event carries the past and structures what is about to occur (PP, 386–387). The expressive qualities of movement that emerge in a film projected at roughly twenty-four frames per second are not found in the individual frames taken in isolation. Those properties (the movement, the narrative, the rhythm, the emotional tonality) arise only at a specific temporal interval, neither below nor above a certain threshold. Ontologically speaking, the cinematic world disclosed in the phenomenal present is not "there" in the frames (distal stimulus): it emerges as the result of a relational process.

The cinematic narrative, what unfolds in the thickness of the phenomenal present, exists only at this level of reality (direct experience) and complexity (the event taken in the fullness of its relational structure). The same applies to our body, our lived experience, and our mental life: these cannot be reduced to the analysis of their physical components. The physiological body does not coincide with the lived body; neural processes do not exhaust the phenomena of thought. The fear one experiences during the shower scene in *Psycho* is not contained in the individual frames of the film strip. Although they represent part of the conditions of possibility for the emergence of the film, the two planes, the film and the frames, do not have the same characteristics. It is time, a precise relation, that brings forth the movement, the expressivity, and the spatiotemporal flow of the cinematic event, that *continuum* which constitutes the thickness of the present.

If we were to treat cinema as an illusion, we would have to treat all movement as illusory, since all movement is, in a sense, stroboscopic (SW, 58, 150). But then where would reality be? Calling something an illusion presupposes the existence of another domain that is "real." Yet the distinction between appearance and reality is not ontological but epistemological: it depends on the frame of reference through which we judge the world.

Thus we call the "bent stick in water" an illusion relative to the "straight stick out of water"; we call virtual reality "apparent" relative to the actual world; we call the perceived world "apparent" relative to the physicist's world; we call the continuity of cinema "apparent" relative to the static frames; we call the Sun's movement "apparent" relative to Earth's rotation; and we call the fused grey of a spinning disk

[2] Cf. P. Bozzi, *Vedere come. Commenti ai paragrafi 1–29 delle «Osservazioni sulla filosofia della psicologia» di Wittgenstein*, Milano: Guerini e Associati, 1998, pp. 45–46.

"apparent" relative to the stationary disk.[3] Yet each of these distinctions reverses as soon as we adopt another frame of reference. What had previously been taken as mere appearance or illusion may in turn become the reference reality. One can sleep and dream within a virtual environment. Within that dimension, the dream counts as appearance relative to the "reality" of the virtual world inhabited by the sleeper, while this virtual world itself appears as mere appearance when measured against the world in which we are now situated. Likewise, the table perceived in virtual reality may be said to be "apparent" relative to the table present here in this world; yet this latter table, too, becomes an appearance when viewed from the standpoint of the physicist's world, where it is reduced to a set of relations or particles and understood as dependent upon our perceptual apparatus.

The experiments conducted by Gelb in 1929 exhibit the same reversibility. A dark grey disk placed under a spotlight appears white; when a small piece of white paper is placed upon it, the disk again appears grey.[4] If we spin a half-white, half-black disk until it produces the same grey, the question arises: which grey is real? Is the grey disk real in relation to the white one, or is the disk that is partly black and partly white real in relation to the grey one? On what assumption (on what frame of reference) do we judge one to be apparent and the other real?

The chromatic fusion is, on the color plane, the analogue of the temporal threshold in cinema: at a certain point, a quantitative relation gives rise to a *qualitatively* different emergent property. The same logic underlies figural pregnance, water boiling at one hundred degrees, and the projected movement of a film: emergent qualities belong to relational structures rather than to underlying levels.[5]

In the world we encounter a multiplicity of aspects that do not exist (or emerge) at the level of the distal stimulus, and conversely there are features that exist at the level of stimulation yet remain imperceptible: classical cases are those of masking or mimicry. From the phenomenological point of view, everything that appears lies on one plane of immanence. Virtual worlds, too, belong to appearing. Illusions are part of the visible world, just like colors, shadows, reflections, and digital images. Different worlds, but a single reality, intersect through appearing. Bees, for example, can distinguish a Picasso from a Monet: they share the Gestalt of the paintings, even though the meaning differs for them.[6] Each form of life orients its behaviour within its *Umwelt* according to meaningful structures. Merleau-Ponty's reflections on nature thus open onto an ontology of relations.

The link between temporality and meaning, time being "literally the tenor of our life" (PP, 500), expresses our belonging to the world as a form of life (the lived body)

---

[3] Cf. T. Agostini, *Il colore acromatico*, in F. Purghé, N. Stucchi, A. Olivero (eds.), *La percezione visiva*, Torino: Utet, 1999, p. 474.

[4] Cf. M. Massironi, *Fenomenologia della percezione visiva*, p. 179.

[5] In this sense, Gestalt theory belongs fully to the paradigm of complexity: a *Gestalt* cannot be reduced to its parts; it emerges as a dynamic, autopoietic form from their *interaction*.

[6] Cf. G. Vallortigara, *The Origins of Consciousness: Thoughts of the Crooked-Headed Fly*, New York and London: Routledge 2025, Chap. 7.

that is situated, open, and oriented in a certain "direction of meaning." Merleau-Ponty's analyses do not merely repeat what has been said about the world; they illuminate afresh the sense in which "subject and object as two abstract 'moments' of a unique structure which is *presence*" (PP, 500). Whatever meaning we assign to "meaning," we always encounter the opening (*ek-stasis*) of subjectivity toward the world: our incarnate being is always already "projected" toward something other than itself. "The world is inseparable from the subject, but from a subject which is nothing but a project of the world, and the subject is inseparable from the world, but from a world which the subject itself projects" (PP, 499–500).

Chapter 19
The New Flesh and Technological Embodiment

To thematize the blind spot of perception, or the impossibility of directly perceiving one's own eye during vision, concerns aspects that do not belong to perception in the strict sense: we experience things directly; we do not perceive our perception of them, nor do we see the physiological processes that underlie perceptual activity. However detailed a physiological account of the hand may be, it tells us nothing of the meaning of a gesture, whose sense belongs instead to the phenomenological register. Similar considerations apply to the technological devices we use daily: their meaning for us is autonomous with respect to the hardware layer that supports them. Our relation to such devices is constituted by the interfaces we manipulate, not by the functioning, unknown to most, of the underlying mechanisms. Interface design in computing must be consistent with technological development, yet is also logically independent from it, because it answers to the bodily and spatial schemas of use.

Since the relation between subject and world is not *directly* given, it must be approached also in conceptual and ontological terms. The perceptual field is a nexus of relations, but what is given to us are only their emergent qualitative and expressive effects: an invisible-visible relational chiasm. Relation and perception are not themselves *objects* of experience: "perception is nowhere" (PP, 44), and must not be reified; relation is *constitutive* of perception and can exist in itself in order to make other things exist (PP, 44). To grasp the meaning of the world's appearing requires intersecting the epistemological and ontological registers: understanding the modes and terms of relation (ontology) while identifying (epistemology), through experimental phenomenology, the dependent and independent variables, that is, the conditions for the emergence of a phenomenon. As in the study of movement (stroboscopic movement and the phi-phenomenon), appearance arises only under determinate conditions (ISI), from specific numerical ratios and not from others. The analysis of such conditions reveals–this is its speculative significance–the chiasm between ontology and epistemology, without reducing one to the other.

Analogous "factors" can be observed across sensory modalities: visual, acoustic, tactile. A well-known case is Wertheimer's factor of proximity governing the

L. Taddio, *The Appearing of Sense*, Lecture Notes in Morphogenesis,
https://doi.org/10.1007/978-3-032-27021-4_19

grouping of perceptual units. The unity of a thing at the phenomenal level enjoys a certain independence from the physical level of investigation: Gestalt unity occurs within the world on the basis of relations that emerge only at a given degree of complexity. Phenomenological analysis must therefore adopt the appropriate conceptual framework for the level of reality under consideration: this is the essence of the phenomenological relativistic attitude.

One should not expect to find in physical analysis the same features discovered at the phenomenal level, for each concerns a *different* conceptual plane. What must be clarified each time is the way in which the epistemological plane intersects with the ontological one, without absolutizing either. Philosophy is often tempted to absolutize *one* theoretical perspective; this is why the careful examination of individual cases is so revealing, for it exposes *their* perceptual and conceptual grammar. The phenomenology of perception investigates the world, whereas the underlying physical reality, being inaccessible as such, requires other tools and methodologies. Recognizing the different (epistemological and ontological) levels of complexity of reality entails a *relativistic* method in phenomenology, which is not arbitrary but conceptually disciplined. Clarifying the *frame of reference* is one of its primary tasks.

To describe what we see is to remain anchored to the perceptual field, which may be treated as a plane of immanence for experience, the proper object of a *science of observables in act*. That reality is not directly accessible does not entail that it cannot be known, nor that subjectivity constitutes an insurmountable limit to knowledge. It is insufficient to assert that because there is an experiencing subject, knowledge must be mediated and therefore distorted. This mediation must instead be delimited, clarified, and examined in its phenomenological implications.

We see the interlacing of worlds and reality when we observe the behavior of other living beings. As ethologists show, animal behavior provides an index of "reality": from reality emerges the coherence of behaviors across a plurality of worlds corresponding to different forms of life. As Lorenz notes, these systems are not mutually contradictory and do not exchange incoherent information. In our world we observe the behavior of beings inhabiting, through their bodies, a different world, one shaped by a different image-apparatus. We also investigate reality through microscopes or telescopes, where interpretation plays a necessary role: a different conceptual grammar is enacted, and the system of relations shifts because we interpret what we see, implicitly taking the instrument as an extension of the body that reveals an otherwise inaccessible plane of reality. As noted earlier, we do not see a micron under the microscope; we see an image several centimeters wide that we *know* to be a micron. Likewise, we see a dot or a yellow sphere that we know to be a light-years away, though we do not directly perceive this distance. These examples illuminate Merleau-Ponty's claim that "wild" Being is the "common tissue of which we are made" (VI, 203), describable only "as that pre-spiritual milieu without which nothing is thinkable, not even the spirit" (VI, 204).

Phenomenology does not seek to establish a universally valid epistemological framework, but to clarify, each time, the conceptual frame appropriate for interrogating a given domain of knowledge. The supposed short circuit between language and perception, and the alleged impossibility of surpassing them to attain direct access to the sensible, are epistemological assumptions that overlook the ontological significance of the *relation* binding thought, language, and world. Can thought emerge independently of any experience of the world? Understanding this relation requires adopting a paradigm of complexity, grasping emergent properties arising from relational dynamics. The notion of *Gestalt* is exemplary in this sense.

The expressiveness of the world, things as they appear and begin to disclose themselves through their evocative force, exists only by virtue of the relation between subjectivity, body, and that correlate we call "world." We are immersed in relations that constitute the vital core of existence. Silence, for Merleau-Ponty, is the dimension that precedes speech; phenomenologically, this also means acknowledging the impossibility of experiencing absolute nothingness or absolute silence. Silence, relative to speech, functions as the background of a figure. We always perceive something in relation to something else. Even the deepest darkness is still "something"; the most absolute silence can become "deafening." Our implicit frame of reference (the body) cannot be detached from what constitutes it, namely the world. Likewise, the world, as the correlate of experience, would not exist as we know it without a form of life capable of perceiving it. To describe the world is to embrace every mode of its appearing as an expression of the virtuality of being: a reflection is no less real than the sun that makes it possible; a shadow is no less real than the object that casts it. The same grammar, the same subtle play between appearance and reality, is at work in Merleau-Ponty's analyses of illusions. Rewriting the prose of the phenomenal world through description is his way of allowing becoming to live in language, of making words vibrate with the expressivity of the visible.

We are woven into this "wild being": "All this issues in a theory of perception-imperception, and of the *Logos endiathetos* (of meaning before logic)" (VI, 168–169). Things possess a meaning embodied in the world, given in a play of "reversibility" that is neither subjective nor objective; it emerges, as in Gibson's affordances, from relational properties. Meaning and world can only be justified as "aesthetic phenomena" (Nietzsche). "There is no intelligible world, *there is* the sensible world" (VI, 214) for every form of life. From this perspective, the study of illusions has enabled us to reconsider the relation between appearance and reality in terms of reversibility (epistemology), and to understand the world itself—as "rays of past/of world" (VI, 240)— as a *virtuality* of reality (ontology): a "wild being," raw and impure, eternal and immobile, *principium individuationis*.

Chapter 20
Conclusions: The Virtualities of Being

I.

The concept of *reversibility* is developed by Merleau-Ponty to grasp our gaze as a perspective that, in "mirroring" itself in the world, returns to us a dimension already imbued with meaning. If, for Lacan, the mirror stage is constitutive of subjectivity, for Merleau-Ponty the world itself functions as a mirror insofar as it envelops, reflects, comprehends, and ultimately constitutes the meaning of our actions, granting significance to our very subjectivity as a relational event. The body too, understood as *Gestalt*, is a perceptual system embedded in this "open" circularity sustained by field dynamics. Our corporeality lives in the chiasm: we are simultaneously seeing and visible, desiring and the object of desire, active and passive; "every relation with being is *simultaneously* taking and a being taken" (VI, 266). Inside and outside form a kind of "membrane" that filters information like one organism within another: forms nested within forms. We are not a closed and stable system, but a processual, metastable, open, and stratified one.

We can understand different levels of reality in relational terms, even though each level of complexity requires its own conceptual tools. The transformations that accompany the becoming of things concern even the concept of "death," understood as the collapse of our specific form of life. It is within this flow of self-generating forms that we can perceive the break with classical metaphysics, which sought to ground truth in the need to affirm a stable meaning for human existence (*episteme*). Philosophy, religion, and art may then be seen as epistemic articulations of our form of life: without the human being, there would be neither God (religion), nor truth (philosophy), nor beauty (art), nor even this world as such. Meaning unfolds within the world and through its justification, beginning from our corporeality. In this sense, the phenomenology of perception also reveals an existential dimension: "everything speaks to our body and to our life" (C, III, § 4; my translation).

For Merleau-Ponty, the painter has the prerogative of observing the world without being obliged to judge it. This marks a crucial difference from our habitual relation to things. Everyday evaluation accompanies and channels our perception, orienting it

L. Taddio, *The Appearing of Sense*, Lecture Notes in Morphogenesis,
https://doi.org/10.1007/978-3-032-27021-4_20

toward use. Things appear as manipulable instruments within technical and productive practices; our relation to the "thing" is tacitly bound to technique as a "means" for achieving an "end." When we see a train in a station, for instance, it appears to us primarily as a means of transport. Heidegger, in his reflections on the thing, later reinterpreted by Merleau-Ponty, recognized this essential trait of the metaphysical tradition: the reduction of being to manipulability, culminating in the technological age. Merleau-Ponty's point is that if we relate to things exclusively through use, we no longer see them as they truly appear.

René Magritte pursues a parallel aim within his own artistic practice: restoring things to their full visibility, showing them as they appear in truth. His method seeks to reveal our original relation to the visible. Ordinarily, we do not look at a train for how it appears; we see it through its utilitarian meaning. In *La durée poignardée*, Magritte places a locomotive emerging from a fireplace inside a domestic room, creating a visual shock that suspends, even if only for a moment, the everyday meaning of the object. This temporary suspension–an implicit *epoché*—allows us to encounter the locomotive in relative isolation from its utilitarian context. This is precisely Merleau-Ponty's insight: painters can teach us how to see the world. It is on this phenomenological basis that his philosophical interest turns to painting, culminating in his sustained engagement with Cézanne. Painting, for Cézanne, demands an almost ascetic attentiveness: "one hundred working sessions for a still life, one hundred and fifty sittings for a portrait" (SNS, 9).

The painter does not judge the world but, through artistic practice (and the affordances evoked by art), establishes values by revealing intensities and the non-homogeneous structure of the field of existence. The world, from its earliest manifestations, carries implicit values in the *orientations* of our behavior: meaning emerges not from an intrinsic rationality of the human being but from the *reversibility* of our form of life in relation to the world. Meaning thus implies *directionality*, orientation, and a fundamental openness. The system of relations that structures the appearance of things corresponds to this first articulation of meaning, what Husserl called the "*logos* of the aesthetic world." Merleau-Ponty radicalizes this notion by fully situating it in the world: art reveals this dimension of the logos because we are incarnate.

II.

Gibson coined the term *affordances* from the verb *to afford*, meaning "to offer" or "to make available," referring to an animal's "action potentials," the actions made possible by the interplay between the characteristics of the environment and the body (its "form of life"). Objects sized to fit our hands are "graspable": a relation of *reciprocity* and *complementarity* between body and world makes certain actions possible. A form of life is integrated into its environment, natural or digital. The body becomes a *measure* of the world, enabling an evaluation governed by values implicit in perception. Affordances correspond to expressive properties directly perceived: neither subjective nor objective, but relational, rooted in the lived body (*Leib*) and the environment. They also evoke behavior on the basis of the meaningful charges embodied in forms.

Art does not imitate; it recreates resemblances as "forms of equivalence" with the world. It recalls the world's internal morphogenesis, animated by the same meaningful charges linked to our bodily frame of reference. These are worlds capable of evoking articulations of meaning that cannot exceed the phenomenal, and which—at their most radical—seek to expose the "transcendental form." Contemporary art expresses this "impossible gesture": striving to speak beyond itself, revealing a "pre-human" dimension and exposing the conditions of its own making. Yet no gesture can coincide with this originary dimension; it becomes "other" at the moment it appears. Every gesture falls back into the meaning proper to our form of life. Even a monochrome canvas returns to this: art is "conceptual," arising from the impossibility of rendering a "transcendent" or "auratic" truth. It points instead to the forms, processes, and meanings of becoming. Cézanne does not freeze an instant but seeks to express its genesis. Merleau-Ponty first sees this in Cézanne and later in Klee: the values a work evokes emerge from the continuous transformation within the pictorial process. Painting makes the invisible visible by creating affordances (expressions) within processual virtuality.

Reversibility entails that the world "looks at us." The painter embodies this by lending their body to the world (through reversibility, or chiasm) so that it may become painting. The mirror thus becomes emblematic of this relation. The notion of "field" already expresses an "intensity" of the physical world, an intensity also found in the canvases of Rothko. Yet art can express the physical only through a meaningful order that is irreducible to physical reality as such: we remain within the *il y a*, the preliminary "wild" dimension. If we accept the distinction between reality and world, we can see why there can be no art of the after-human, no art without a subject. Art is the expression of *our* form of life. It reveals how a people or culture has given *form* to the "mystery" of creation and to its being-in-the-world.

Art, like the world, produces meaning. Congo, the gorilla studied by Morris, may display aesthetic sensitivity to form and color, but his drawings become artworks only within a human world, under a human gaze. They show aesthetic sensitivity (to forms and colors), not artistic meaning. Today algorithms can generate works indistinguishable from human ones, yet such objects gain artistic value only relative to our form of life.

The centrality of the body and of *flesh* in Merleau-Ponty's philosophy illuminates the relation between subjectivity and technology, revealing how technological development brings us to the threshold of an anthropological mutation, between human and post-human. This is part of Merleau-Ponty's legacy: the theme of the "new flesh," to borrow from David Cronenberg. Here, in conclusion, we gesture toward some implications.

It is not art, religion, or philosophy that "die," but our subjectivity as we have known it, which may be transformed by the rise of technological apparatuses and the progressive incorporation of new technologies. The hybridization of body and technology marks a limit beyond which our form of life can no longer be what it was. No matter how many organs we replace or capacities we enhance, we cannot indefinitely modify the body without turning it into something other than what it is. Altering a form of life modifies its relations with the environment, its needs, and its ways of embodying meaning. Humans have historically found stable meaning through religion, art, and philosophy—distinct modes of adhesion to the world through spirituality (religion), rationality and "truth" (philosophy), or the expression of the mystery of existence (art). Technology, rooted in science and arising from philosophy, now reshapes these relations.

The stability of meaning orients us within the world from the outset. Art reveals the expressiveness of worlds of which the body is the manifestation: this is the sense of Merleau-Ponty's equivalence between art and world in terms of reversibility. Each form of life confers meaning in relation to *its* environment. Art, religion, and philosophy fade as the order of meaning constituting our form of life changes. As our form of life transforms, so too do our ways of inhabiting and embodying the world.

Consider once more an example: in chess, the knight's move has value only within the game; the painter's gesture only within painting. Without subjectivity, the knight is just a wooden piece; the brushstroke is merely pigment, a material object without value (meaning). The difference between world and reality mirrors the difference between the wooden piece and its meaning as a knight. This may seem to echo postmodern arbitrariness, but the view is quite different. Learning chess means recognizing that moves are not equivalent: their value differs in the dynamic of play. There is no transcendent "perfect game", yet moves are not arbitrary. One cannot play at will without destroying the game itself. Meaning emerges as the game unfolds, once symmetry is broken, through reciprocal moves and evolving relations.

The same applies to artistic creation: there is no right or wrong gesture before the process begins. The homogeneous blank surface must be broken so that a dynamic can arise from which norms emerge.[1] We may play chess against a machine or generate art through algorithms, but in these cases there is no meaning in the action: it is pure calculation. Machine–machine interaction lacks meaning, because it lacks the analogical dimension. Meaning arises for humans in *playing*, just as in art. One might object that chess is a closed system with finite moves, unlike the open dimensions of art or ethics. But the objection is abstract: though chess is finite, we could never exhaust its possibilities within a lifetime; relative to us, it has an "infinite" character. Ethics and aesthetics also take shape within semi-closed systems: communities, cultural forms, practices. And every form of life tends toward self-preservation, which can already be understood as an "end".[2] From these premises, one might even imagine dynamic relational matrices—emergent structures of order and significance—as a form of "intelligence" inherent to matter, an informational potential (*bit*) expressed in and through relations.

III.

The distinction between "consciousness" and "object" drawn in the *Phenomenology of Perception* makes it impossible to understand an "objective" fact such as a brain

---

[1] See L. Taddio, *La misura dell'abitare. Osservazioni sull'estetica di Wittgenstein*, "Estetica. Studi e ricerche," 2/2019, pp. 521–538.

[2] Parenthetically, our *interactions* are also always expressions of relationships that determine non-arbitrary emergent properties.

lesion that produces a "disturbance of the relation with the world" (cf. VI, 200), as if consciousness functioned independently of the body. This objective conditioning depends on an event inscribed in the "verticality" of being, not in our conscious life. Once again, the ontological and epistemological planes intersect. The shift in perspective is radical: from the analysis of the structure of behavior, to that of our conscious life in the phenomenology of perception, and then to the "flesh of being," that is, the investigation of the "vertical plane" (reality). These are epistemological perspectives that interweave, revealing different nuances and, through their multiplicity, capturing the *ambiguity* of certain phenomena. To move from the horizontal plane to the vertical plane is to move from what we have described as the "plane of immanence" (the plane of phenomenological experience) to the vertical ontological plane, the wild being in which we are woven and in which we participate (the depth of Being). In the *Phenomenology of Perception* the term "immanence" is used in contrast to "transcendence": the former indicates the activity of thought, the latter the possibility of going beyond it in order to rediscover that original contact with the world which is the object of phenomenological inquiry. Yet Merleau-Ponty himself relaxes this classical conceptual "opposition", or rather grasps it in its constitutive ambiguity, beyond any dualism, when he affirms that whenever one term of the relation is given, the other necessarily appears; the opposites find a balance and each requires the other. There is no transcendence without immanence and no immanence without transcendence, and "this ambiguity is not some imperfection of consciousness or of existence, but the definition of them" (PP, 387).

We must therefore analyze the scope of the reference systems implicit in the relation between transcendence and immanence. Let us first consider the *cogito*, within our relational framework, as the articulation of an originally sensible and non–self-sufficient dimension (it is inconceivable to imagine a form of self-conscious life entirely devoid of perception). In this sense, transcendence always falls back into the plane of immanence, beyond the polarity of inner and outer experience. One transcends the perimeter of the *cogito*, of consciousness, but not that of phenomenical experience, which includes the entire range of internal and external phenomena within the relation or structure of world and lived body, that is, subjectivity. The excess of the visible and the articulation of visible and invisible are no exception: they always remain part of experience, as we see in the analysis of the figure–ground relation or in cases of amodal completion. Strictly speaking, this play of conceptual references may appear superfluous insofar as it is fully absorbed by an accurate description of the phenomena themselves.

The plane of Being encompasses and envelops the plane of experience. The relations that determine the emergence of meaning from the structure of world and lived body are a "layer" of Being: our form of life unfolds within this plane (world), which is only one among multiple planes (layers) of Being. Each form of life expresses and institutes new relations and their corresponding emergent properties, a world interwoven with the same fabric of reality in which we are ourselves enveloped and in which we participate. "Flesh" names precisely this fabric of being, or "reality," in our terms. Subjectivity resembles a wave that preserves its singularity and specificity when observed closely, while, when the perspective is widened, it dissolves into the sea of its becoming (N, 120). Each part, considered in isolation, maintains its specificity, yet the functions emerging at higher levels of relation possess their own singularity. This applies to organisms, to things, and to the complexes of forms nested within other forms that compose the ecological system. The same logic may be extended to social and political organizations, which can be regarded as complex "organisms."

It is this morphogenetic dimension proper to brute or wild Being that constitutes the textures which envelop and encompass our lifeworld, the very world that, from our perspective, is an "opening," an access point for questioning the transphenomenal real. In this sense, Merleau-Ponty articulates the notion of the truth of perception, that is, of participation in Being: if we were not already, in some way, inscribed within truth, how could we ever gain access to it?

The world is a relatively stable plane of Being (we do not directly perceive the continuous process at work), a plane in which the principle of non-contradiction holds and prevails, well suited to the relative stability of phenomena (*principium firmissimum*), and which underwrites the classical opposition between being and becoming. To grasp Being as process—that is, the co-belonging of being and becoming, and becoming as "generative virtuality"—to apprehend the process of transformation beyond this apparent "metastability" that we call "world," requires, as Simondon has clearly shown, an inquiry into the pre-individual, into the *principium individuationis*. It is precisely this vertical dimension that Merleau-Ponty explores in order to understand our relation to things and to the world as a co-belonging to the common fabric of relations, the "flesh": forms within forms in a process of continuous transformation.

Every "thing" and every subjectivity (the lived body) is only apparently stable, that is, metastable. The process itself is not directly visible; the world is a necessary illusion for adaptation to a given environment, a dynamic and expressive field of existence that guides and encompasses our action. It is this meaningfulness that makes possible the emergent properties arising from our activity, from our individual and social (intersubjective) actions. Social organization, as well as the technical transformation of the environment, are always metastable (morphogenetic) forms, that is, virtualizations of the same Being. In this sense, today's "virtual reality" is real: it is an extension and prolongation of the real, whereas artificial intelligence, at least for now, is not an integrated and embodied system within the environment.

To problematize wild Being, the analysis of appearance alone no longer suffices: the investigation of the real requires the support of other sciences such as physics, chemistry, biology, and cybernetics. Ontological inquiry can only proceed indirectly, by working within the spaces of conjunction among the various regional ontologies. Following the thread of the lectures on nature (delivered at the Collège de France between 1956 and 1960), one sees how Merleau-Ponty explored different disciplines—contemporary physics (in the notion of "thing"), biology (in the transformation of forms and evolutionary theory), cybernetics, and others—in order to identify paradigmatic examples from which ontological and epistemological implications could be drawn. This mode of inquiry, already adopted in relation to Gestalt psychology and more generally in dialogue with the human sciences (linguistics, sociology, psychoanalysis), must be assumed as a phenomenological method: a way of working within and through the sciences, a philosophy *in science* and not simply *of science.*

Philosophy takes shape in the *intensity* of its gesture. The transversal cut carried out across different domains of knowledge is dictated by the very nature of the problem and by the degree of intensity with which it is pursued. It is not "imposed" by philosophy (which is not a discipline), but emerges *through* the critical analysis of the knowledge of its time. Science does not require legitimation from philosophy; yet, insofar as it entails an experience, it becomes an object of phenomenological inquiry. Philosophy, in turn, must be understood as the unveiling of the pre-theoretical dimension. Science, too, bears a "relation to being," which opens itself to science just as it does to "individual life" (RC, 75). From this perspective, we can revisit our subjectivity and our need for "meaning" in relation to that operation of synthesis of knowledge which has always characterized philosophy. Even if we consider it misguided to try to reintegrate the subject directly into scientific practices in order to grasp the meaning that such knowledge holds for us, there nonetheless remains the task of investigating our need for meaning as such, a need that is constitutive of existence not only at the individual level but also, and above all, at the intersubjective level.

Both reality and appearance belong to Being, and both can be described as systems of relations. Being is, in this sense, "articulation", "infrastructure", "hinge", apprehended at all its levels and not as something lying behind appearance (N, 236–237). Phenomenological analysis has led us to reconsider the system of relations implicit in the opposition between appearance and reality, no longer treating it in ontological, dichotomous terms as an "objective reality" opposed to appearance, but rather as an epistemological "reversibility" grounded in a single reality. The phenomenon as such is neither object nor subject (RC, 132). The being of the thing manifests itself in a topological space internal to Being, in a multiplicity of virtualities that concern both the thing and our body (body schema). Every alteration, transformation, or change participates in the same Being: there is no ontological "dualism." Technology, too, must be understood in purely relational terms, as an element of transformation.

We have thus rediscovered the things themselves as cores of relations. From an ontological point of view, things and substances reveal themselves as processes, events, and variations of differing intensity. From an epistemological point of view, through the methodological relativism we have adopted, it becomes evident that the internal relations of the perceptual field ground the appearing of the thing, grasped in its relative stability. These processes of transformation are virtualities of the same Being. Consequently, between world and reality there is no simple dichotomy, but rather chiasms and thickenings tied to the multiplicity of relational play. To reflect on the virtual is to understand, in an ontological sense, the meaning of relation. "Man is but a network of relations" (PP, 530). The body, the thing, and the world are virtualities, "rays of world." Merleau-Ponty's expression indicates a segregation, a relative isolation, in a metastable sense. The possibility of a gap, of a reflection on the two (VI, 218), implies grasping the very meaning of bifurcations in non-linear processes. This working hypothesis on the virtual, understood as the vanishing point of Merleau-Ponty's relational ontology, can open new lines of research, from the analogical to the digital, from the human to the posthuman.

IV.

What consciousness does not see is "what makes it see, is its tie to Being, is its corporeity [...] is the flesh wherein the *ob*ject is born" (VI, 248). The *punctum caecum* of consciousness marks both the point at which the retinal fibres are activated to make vision possible and the structural nucleus of invisibility already encountered in the analysis of the appearing of the thing. What we cannot see directly is precisely that invisible core which nonetheless makes our conscious life possible. Visual perception cannot be exhausted by the physiological analysis of the eye, and the same holds for our body. In perception, every bodily stance in relation to the world corresponds to a perspectival "gaze" upon the thing; this bodily disposition is tied to a horizon of meaning that determines our possible actions in space, the virtuality of acting. Both seeing and the sense of our body are bound up with action: they depend on what our body *can do* rather than on what it "is." "The meaning of the gesture is not contained in it like some physical or physiological phenomenon" (PP, 225). It is precisely this possibility that inscribes perception within action: our *capacity* to explore the world, to take a direction, and to interact with it together with other subjects and other forms of life, interwoven within the same fabric of Being.

The world has "meaning" for us. The body that inhabits it does not experience the "thing itself," its "negation," the object "a" *being there* at the centre of Holbein the Younger's *The Ambassadors* (1533). To recall the example used earlier, for Lacan this would imply a vision of the "negative," the instability of the "thing" beyond "representation." In Nietzschean terms, that thing possesses its stability, its "value," its "meaning" only within a "perspectival function." If the "meaning" of the "thing" presupposes a certain degree of "stability" of the phenomenon, then we may say with Nietzsche that "only as an aesthetic phenomenon is existence and the world eternally justified." The "logic of meaning" presupposes a relative stability. Without it, we could neither "name" nor "think" nor "use" a given thing.

The "world" is an aesthetic phenomenon arising from its relation to a given form of life, a structure that does not exist without the other term of the relation, the "subject." To ask what the world is without the subject is as meaningless as trying to understand the fluidity of water by considering only its "isolated" components, oxygen and hydrogen. The fluidity, clarity, and transparency of water are emergent characteristics of a specific structure of the real (H_2O) and cannot be reduced to the elements that compose it. The fact that "something" appears as water does not suffice to make it water. This can be established only through the analysis of reality and not directly from the standpoint of the world, even though, for practical purposes linked to our adaptation to the environment, our visual, tactile, and gustatory perceptions are sufficient. Moreover, if something were to appear as water while in fact being another substance, at the phenomenological level of direct experience our description of the substance and of its resemblance to water would remain unchanged. Once again, this depends on the reference system adopted and on the way we bring the world of experience into relation with the categories and models used to investigate reality. Since the world is a part of reality conveyed through our subjectivity, the relativization of the world is not contingent but necessary for knowledge of the real itself. Otherwise, only the "paradoxes" and "aporiae" produced by attempts to explain reality from a single perspectival plane can emerge. We must not think, as Gilbert Simondon argues, of "relation" as a mere connection between two pre-existing terms, but as a process that takes shape on the basis of the state of a system.

Perception, as Merleau-Ponty teaches, is the outcome of a complex process that extends from the stimulus to perception itself, understood as the "conscious" result of that process.[3] Consciousness is not the final endpoint of the chain, for a *new* exchange of information begins from it and in turn feeds back into the body: there is reversibility. As in a Möbius strip, inside and outside intertwine and flow into one another. It is at this level that we can situate a new kind of *relational realism*. Realism should be attributed to our way of conceiving "relation," which is not the outcome of the two isolated terms of the relation "SrO", where "S" (subject) and "O" (object) are conceived as self-standing and only later as elements that give rise to the relation. It is not the terms "S" and "O" taken separately that generate "r." The analysis must be inverted: it is "r" that determines the existence of "S" and "O". The relation "SrO" must be understood as a "structure," a *Gestalt*. The emergent properties we experience arise from the structure. The "Kanizsa triangle" is the result of such self-organizing properties of the field. In this sense, the experienced "world" is neither "arbitrary," nor a "construction" of the subject, nor a mere elaboration of our brain. The triangle exists, *it is there*, and it is in the world that we *locate* it.

One might object that the "triangle" does not exist in the physical world, at the level of the distal stimulus. Yet we know the physical world only through the phenomenal world, the world we *encounter*. Moreover, in scientific terms we are prepared to define as "objective" a phenomenon that is intersubjective and repeatable. An optical illusion satisfies precisely these criteria. Things that appear, in their different modalities, inscribe themselves within the world; they are part of our environment just as films, illusions, reflections, shadows, trees, and mountains are. They are never isolated; they presuppose, as we have already seen, at least their constitutive relation with a background, as shown by Wertheimer. Things possess relational properties and exist in the sense that we can experience them, enter into relation with them, and interact with them. Our structural bond with the world is not immediately evident but calls for careful phenomenological analysis. In naïve contact with the world, we do not perceive any strict, direct dependence between ourselves and it. This is why, at first, we are all "naïve realists": we feel free and autonomous in relation to it. The things we encounter in the world possess different degrees of independence, stability, and resistance to change. It should also be noted that even the objects we see and manipulate in immersive worlds must display the same phenomenal characteristics—namely, a relative independence. The nucleus of relations that constitutes both the

[3] The very connection between perception and consciousness should be reconsidered in light of the experiments conducted, among others, by Vallortigara (*The Origins of Consciousness: Thoughts of the Crooked-Headed Fly*), which show that patients with lesions in the visual cortex demonstrate residual visual capacities and "inferential logical" abilities below the level of full awareness.

natural and the digital thing unfolds through a game of relations between the visible and the invisible. Otherwise, we would not even be able to experience it.

V.

Every *form of life* participates in a single reality and adapts to it through the relation and interaction of the body with the environment. The world emerges *with* the phenomenal properties we know through immediate experience. If the world were not the result of a physical process, the underlying reality would remain unknowable: we gain access to reality only by starting from the investigation of the "world-environment," that is, from direct experience, which we see and live, since there is a certain degree of coherence, a monism, between "world" and "reality." If we were to assume the *cogito* as the sole foundational moment of knowledge, without considering it as the outcome of a relation with the world, reality would remain closed to us. Like Descartes, we too would need a God to restore our original contact with the world. We cannot imagine any form of thought that is not coordinated with a world: thought is always a second expression of the world. Thought is never perfectly autonomous, and prolonged sensory deprivation in fact leads to forms of cognitive regression. If the world has meaning, it is because subject and object are aspects of one and the same structure, a *Gestalt*. Otherwise, the concept of "reversibility" would collapse and every phenomenon of "feedback" would be exhausted, lacking a relational structure.

Can we conceive the emergence of a "state of consciousness" from a body entirely deprived of sensory perception of the surrounding world? If the answer, as we believe, is negative, we can transpose the same thought experiment into the digital domain: can we conceive the emergence of a form of artificial intelligence within a system that is not integrated, a disembodied system? Faced with the appearance of a perfectly rational intelligence, deprived of emotions and sensations, of every element of "irrationality," one decisive question remains suspended: would such an entity "want" to live? Life is not the expression of a rational choice; the body "thinks," it is oriented toward existence. The perceptual field is organized and imbued with meaning. For a perfectly rational entity, every direction would have the same value, and for that very reason the choice of one direction rather than another would lose its meaning. The body, by contrast, is alive and oriented to and within the world. It is traversed by drives and stimuli that are functional to its survival. The world speaks to us because we ourselves are in the form of world. In this sense, things, as affordances, are charged with meaning. If we imagine a form of life integrated with the digital, we must find equivalent dynamics: a digital form of life would have to be integrated with its digital environment. If this is the case, our hybridization with technology is no longer contingent. We can manipulate things in a coherent way that is functional to our interaction with the world, extend the virtuality of things, create new worlds, and adapt to them. To do so we can make use of prostheses, although for now the implicit frame of reference remains our body. From this perspective, a modification and redesign of the body become indispensable: the development of the digital entails inclusion in a complex system which, in order to be enhanced, must progressively increase its degree of integration so as to exchange and process more information. However difficult it may be to foresee the future development and integration between ourselves and digital worlds, the absorption of our body into this new world necessarily implies its transformation and enhancement. By hybridizing with technology, we can adapt to the digital world and thereby convey an ever-growing flow of information between ourselves and the digital environments in which we are, and will be, inscribed.

VI.

The framework of existence that emerges from the "necessary illusions," to evoke Nietzsche, must be read in terms of the virtuality of the real, of worlds, and not as a set of "appearances" opposed to the real, "true" reality. There is no opposition between different worlds, since they are internal relations which, kaleidoscopically, belong to the same fabric of Being, a monism. From Being there emerge "appearances," visible and invisible worlds that every form of life requires as elements of a metastability of meaning. Looking, through Merleau-Ponty, to the work of Simondon, we grasp, thanks to the adopted frame of reference and the *principium individuationis*, the processuality of the "world", conceived from the point of view of reality itself and its continual transformations. From a phenomenological perspective, we can adopt for this world the *principium firmissimum*, the principle of non-contradiction, which holds in a relatively stable world, one that is in slow but constant transformation. We do not directly see processes, but only the outcome of processes within our spatio-temporal phenomenal window, which situates us in an internal relation to reality through the world.

The subject inscribes its actions within a dynamic field of existence that changes according to the spatio-temporal coordinates which render its life-cycle relatively stable, allowing it to adapt to the surrounding world, to develop, to reproduce, and to complete the cycle of its existence. If phenomena were to appear within a spatio-temporal relation that was too narrow, that is, in an excessively unstable way, there would be no "thing," no world. Independently of our perceptions and of relation itself—as we have tried to show—there would be no thing: the tree before us would not exist "as it appears," it would not possess those characteristics. As Kant understood, without our presence, without our form of life, there would be no "this tree"; indeed, the very concept of "tree" has this meaning: it denotes an object of the world, not of reality. Kant's "transcendental" could be taken as equivalent to our concept of "form of life," although this concept is not "pure" but must be understood within a morphogenetic horizon that traces the dynamics of the generation of biological forms, in terms of a genesis of the transcendental.

Natural language, phenomenological language, can grasp an object in a non-arbitrary way, just as a poem can convey a certain "meaning," a certain value, and so on, in a non-arbitrary fashion. In the same way, mathematics, as a language, expresses a determinate content. The mathematician may disregard the realism of the world because mathematics also embodies virtuality. It articulates a kind of hyper-realism: if something that does not exist in this reality were to exist in another, it would exist according to the constraints expressed by that mathematical content or that specific theorem. In these terms, we can distinguish the lifeworld, the object of the human sciences; reality, the object of the natural sciences; and the real, the object of mathematical inquiry. Phenomenology investigates both the world and the intertwining of different forms of knowledge. Our world is the object of empirical-phenomenal knowledge: it is knowable as an emergent property of reality, which is itself at the centre of scientific investigation, even though it is not directly experienceable. To analyze reality, science must make use of experiments, models, theories, and the like. Naturally, the scientific method, like mathematics, extends to encompass many of the phenomena that present themselves to our experience. Yet art and religion do not fall within this scope, since they are instituted by human beings in order to express, in a non-arbitrary way—that is, through a technique—the mystery from which the coming-into-presence of forms originates. To explain art, and the same holds for religion, and to lay bare its processes in relation to human needs, means no longer living it within its original function.

VII.

We encounter again the conditions of appearing (the emergence of a given phenomenon) brought to light by experimental phenomenology, even when we move from the perceptual field to a virtual field of perception. Although objects and our body within a virtual space may perform actions that would be impossible in natural space (for example, we may "fly" or encounter things that come to life and transform into fantastic objects) we nevertheless find ourselves in an experiential space. We undergo experiences of things and events that take place and that, in taking place, manifest themselves in accordance with the modes and rules of phenomenal appearing. Since we are dealing with the appearing of a virtual space, it will configure itself, as an object of experience, always as a hybrid of the analog and the digital.

Virtuality does not unfold in "virtual space" as this is commonly understood, in opposition to natural space. Natural space, too, is a *virtuality of Being*, and it is from this conjunction that we can understand why the transformations of the body and its possible hybridizations with technology do not imply any dualism between nature and technique, or between body and technique. The processes of transformation, mutation, and change take place within one and the same fabric of Being.

The constitutive relations that allow a given figure to exist are the same relations that make it impossible for that figure to subsist outside the *nucleus of relations* that constitutes it. By analogy, we can say the same of subjectivity: it can arise only in and through its relation with the surrounding environment. As already noted, world and subjectivity are the poles of a single structure, a *Gestalt*. There is no "thing" or "body" that could appear outside a relational fabric; there are no "things," "bodies," or "individuals" simply "there" *in isolation*. Even the individual, in order truly to be such, must be understood as a "nucleus of relations." If we analyze the relation between appearance and reality within the perceptual field, we never encounter a reality behind appearances, but only appearances themselves. This is why Merleau-Ponty can say that the segments in the Müller-Lyer illusion are neither longer nor shorter. Such attributions, internal to the perceptual field, are evaluative determinations that come after the primordial appearing of the figure. On this basis, we can say that illusion is a virtuality. It occurs in phenomenological immediacy as the lived experience of segments of different length and, as a virtuality (as a system of relations) it is just as real as the number we subsequently read on the ruler. In the same way, the broken stick in water is a virtuality internal to the world of experience and just as real as the "whole" stick, and the virtualities of apparent motion are just as true as any other kind of movement. The values we ascribe to the world thus depend, in a relative yet non-arbitrary way, on the epistemological stance we adopt toward a given domain of our existence.

Consciousness, Merleau-Ponty insists, cannot be fully present to itself at every moment; otherwise, illusion would be impossible. There must be phases in which consciousness does not entirely know what it is doing; if it did, [per evitare ripetizione con "otherwise", poco sopra] it would immediately recognize the illusion as such, and the illusion would thereby disappear (PP, 346). If sensation and perception were given in full simultaneously, no illusion could occur. We may, for example, take a "patch of sunlight" on the road for a stone when the entire perceptual field lends itself to such a confusion (PP, 346). In such a situation we expect a certain kind of tactile resistance underfoot. Yet, as Merleau-Ponty notes, the "correct" perception and the illusory one are not opposed as an adequate and an inadequate way of "thinking" the thing. The ambiguity between what appears and what is real is not localized either in the subject alone or in the object alone; it is distributed across both sides of the relation. Only in this way are myth, dream, and illusion possible (PP, 344). For consciousness, appearance and reality coincide in the sense that whenever we feel, see, or perceive something, this "something" presents itself as certain for us, independently of any external object. Yet there must be some intrinsic feature (PP, 344, 346) that lets us discriminate the specific character of illusions; in other words, we must be able to exclude the idea that perception and illusion share exactly the same mode or structure of appearing. Without such a feature, we could not be conscious of a perception or an illusion as such, nor distinguish the truth of the one from the falsity of the other. There is therefore, for Merleau-Ponty, a structural difference that makes perception, unlike illusion, a true perception. At the same time, for illusion to be lived as illusion, we must "forget" its unreality: the act of consciousness in which the illusion occurs exceeds its own mode of appearing, so that the illusion is not what it seems to be. We cannot, however, cleanly separate appearance from reality within the subject; to do so would cause the very possibility of establishing the truth of phenomena to collapse. And yet there can be error only because there is already truth by reference to which error can be recognized as such (PP, 345). Here, Merleau-Ponty risks a short circuit: when he attributes truth and falsity not only to appearing as a whole ("being" or "living" in truth) but also to singular moments of appearing (illusions), he risks treating illusion as no less true or false than any other appearing, whereas every mode of appearing precedes judgment. To take a perception as deceptive, that is, as illusory, is always to adopt *an epistemological stance* with respect to another perception: it is to compare two aspects of the visible.

Merleau-Ponty repeats that there must be "true" appearances if there are to be false ones, illusions, and that there must be a moment in which consciousness misleads itself if illusions and hallucinations are to exist at all. We disagree with him on this: every form of appearing precedes the division into true and false. It is the judgment that is true or false, not appearing as such; truth and falsity, in the strict sense, belong to epistemology. On the ontological plane of existence we can ascribe truth to appearing as a whole, to appearing understood as openness and belonging to Being. The horizontal plane of immanence is encompassed within the verticality of Being: it is one of the planes, one of the internal relations of Being.

The relation between appearance and reality depends on the frame of reference implicitly assumed through our body, our actions, and the value system at play. To call something false or "merely apparent" is always to do so in relation to another appearing, and this latter, too, is still an appearance, a virtuality of Being. Let us return to some of the cases already examined: the Müller-Lyer illusion, the Kanizsa triangle, the broken stick in water, apparent motion in film, or a fully immersive reality such as in *Matrix*. The Müller-Lyer illusion, like the Zöllner illusion, is paradigmatic because its analysis shows that it cannot in fact be "measured" by any ruler. The system of relations in which alone the effectiveness of the illusion can be judged (the deformation of the two segments within lived space) is given by the variation of the angle in relation to the entire bodily system of *eyes-in-the-head-on-the-body-resting-on-the-ground*, not by the isolated parts of the schema. Reality, opposed to appearance, cannot be defined by simple correspondence to the distal stimulus; it depends on our epistemological stance toward the world. Otherwise, by consistency, we would have to claim that the Kanizsa triangle does not exist, since it has no distal correlate. And yet we see it exactly like any other thing in lived space, according to its own mode of appearing, in this case through ambiguous contours often classified as "quasi-phenomenal," an expression that should be rejected, since the "quasi" already installs a hierarchy and a lack.

The stick that appears bent in water is just as real as the whole stick outside water. In the first case, we receive correct information: the stick is in water and appears according to *that* system of relations. To interpret this as mere "appearance," we have to remove the stick from the water and compare the two situations. But once we do this, the conditions of observability are no longer the same: we are comparing different configurations and declaring one, in relation to the other, to be "mere appearance". One might object that it is the same stick; yet the system of relations that defines the thing changes with the environment. With respect to an observer, the play of relations that emerges in the case of the stick in water is not the same as that of the stick in air. The same holds for cinema and every so-called "apparent" motion: such motion is "apparent" only when judged in relation to another appearing that is taken as "reality." But when we say that cinema is "only an appearance", in relation to what are we speaking? *Where* would the reality of that apparent event we see when we watch a film be found? Would the reel be the "true reality" in relation to the film, just as the black marks would be the true reality in the case of the Kanizsa triangle? Yet ontologically speaking, the cinematic event does not exist in the reel; its plane of existence is irreducible to the physical support.

In *Matrix*, a paradigmatic case of total immersive reality, Morpheus can only reveal to Neo the "real" world by waking him up and leading him into another world. Only from that moment does the previous world *become* an illusion. No proof within Neo's first world can establish its illusory character; every "proof" would itself be another image in the same experiential field, structured by the same bodily frame of reference. Even a photograph of the "real" world would simply be a photograph in Neo's world. Only by "passing" into *another* world, by instituting a new frame of reference, can he reinterpret the first as mere appearance. And yet, if Neo were again to awaken in the first world, it would once more take on the status of reality, and the other would risk being relegated to the status of dream. If a world is coherent in relation to our body, then its reality is *relative*: it is an epistemological stance. To assume it as real or as mere appearance is always done in relation to something else. Its reality depends on the system of relations within which we inscribe our body, which makes possible a comparative judgment between the appearing of two or more worlds. Every appearing is a virtualization of the real: even hallucination, for the subject who lives it, is an undeniable fact. The world that appears within an immersive device is just as real, or just as apparent, as the "world itself." This play of relations, this process of reality, is relative yet never arbitrary, since every virtual plane requires determinate *constraints* in order to emerge.

The "flesh," to use Merleau-Ponty's expression, is the tissue of relations that governs the processes of transformation internal to Being. As the examples of illusion show, these processes do not depend on the subject: we cannot see anything other than what appears under given conditions. The conditions under which a phenomenon emerges are the proper object of an experimental phenomenology, formulated in terms of dependent and independent variables. It is the coherence and non-arbitrariness of the phenomenon that ground our "faith" in the external world (naïve physics). The independence of "phi," the phenomenal event, does not imply any dualism, since *phi* is always embodied in a system of relations, a phenomenological plane of immanence. Flesh names the system in which we are inscribed and enveloped.

The world harbours a system of values and an order of meaning in relation to our form of life; without this relation, the very question of meaning would be empty, since one of the poles of the relation–the subject–would be missing. Reality, as the object of scientific investigation, does not in itself bear meaning; the sciences aim to objectify reality and study it, as far as possible, apart from the implications dictated by subjectivity. To re-problematize this relation is the task of phenomenology: to question the sense that a given body of knowledge may have for us and the effects it may exert on our subjectivity.

The stability of a world, its non-homogeneity and dynamism, are features of the field that determine the directionality of the environment, the folds of Being, and the first systems of value. Values, forces, and orientations of the world are what art reactivates and intensifies. Through art, human beings thicken the world's system of values, endowing it with meaning and with a transcendent justification, an operation as illusory as it is necessary for the survival of our form of life. To move entirely beyond this order of meaning would be to render unthinkable an artificial form of life disembodied from the world, deprived of this drive, of this being an organism integrated into an environment. An artificial subject detached from the world, from emotions, and from the forces that traverse the field of existence, would have no reason to live: as a hypothetically perfectly rational being, it would see all choices as equivalent and thus devoid of value.

Art, religion, and philosophy are the most intense expressions of our form of life: non-human subjects do not articulate their existence through these modes of being and transcendence. If the process of hybridization with technology continues, we may reach a threshold at which the specificity of our form of life is surpassed and, through transformation, another form of life emerges, one that will live according to the dynamics of meaning arising from the relations instituted within its own world and through its unprecedented expressions. Art, religion, and philosophy fade with the human: they correspond to needs inherent to an existence that requires the power of illusion capable of imparting, in a non-arbitrary way, meaning, values, and stability to the lifeworld. The multiple dimensions of sense that sustain the justifications of our existence draw nourishment from the *conatus* that drives forms of life toward self-preservation: an intelligence proper to the body, as Merleau-Ponty teaches, prior to reflective consciousness and also analyzable from an evolutionary perspective. Meaning is a functional illusion of our form of life, orienting it and providing it with a certain degree of stability.

VIII.

The framework outlined through Merleau-Ponty gradually takes shape as a philosophy of complexity and relations, oriented toward describing and understanding the interactions among different levels of complexity within the real, or what he calls wild Being. Its central theoretical focus lies at the intersection of the ontological and epistemological planes: clarifying how these two dimensions intersect, and which implicit frame of reference governs our stance toward the real. The notion of the virtual proves decisive in initiating a first reflection on "flesh" (and, conversely, flesh illuminates the virtual), understood as the relational fabric of Being in which both world and subjectivity are enwrapped: a "layer of being," in Merleau-Ponty's terms, or a plane of immanence, as Deleuze would say.

Living in "virtual worlds" is akin to inhabiting a three-dimensional painting; yet, however many worlds we may create, they all remain inscribed within the same flesh, that is, within the virtuality of Being itself. By lending its body to the world, through the logic of reversibility, subjectivity transforms the world into a virtuality of Being, instituting a new "system of equivalences." The artist, in this sense, filters the world through the body within a dynamic of meaning that is established from the outset through action. The lived body (*Leib*) institutes meaning in the world as part of a single *Gestalt*.

Virtuality thus concerns not only worlds and images but also the hand and its gestures, understood as virtualities of the "body schema" itself. The virtual pertains to the appearing of the thing and to what the body schema organizes in relation to both its visible and invisible dimensions—for example, in the perception of a cube. What we do not directly see nonetheless remains in relation to our body. The phantom limb offers a paradigmatic example of such virtuality. More generally, everything that emerges from relations can be ascribed to the virtual. World and its modes of appearing, the body and the actions it can potentially perform, all belong, on the epistemological plane, to virtuality. From an ontological standpoint, there are no things or substances in themselves, but only events, relations, and processes. Epistemologically, however, we live and grasp reality through a system of metastable relations that we call world and subject—inseparable aspects of the same relational field. The meaning of the world is reversible on the epistemological level: the world and art return our gaze to us, and "it is by lending his body to the world that the artist changes the world into paintings" (EM, 162).

Within the plane of phenomenal immanence, things present themselves exclusively as relations. Scientific inquiry allows us to investigate reality by analyzing forms and processes of formation and individuation across nested systems: from social structures to individuals, from organisms to cells, from cells to atoms. Whereas we once transformed nature primarily to ensure survival, today we reshape the world through digitalization to expand our capacity for transformation, learning, and information exchange. Yet this expansion requires more than a digital medium: our very body must increasingly become an object of transformation and hybridization. Rethinking corporeality and environments thus becomes one of the ultimate challenges of design and architecture and, more broadly, of phenomenology itself—particularly a phenomenology of perception understood in its existential dimension. Against this background, the analysis of appearance, reality, and illusion has shown how meaning emerges as a necessary feature of our form of life. Certain biological and existential traits, such as upright posture, freed hands, and heightened reflexive awareness, have made thought itself possible.

Our incarnated being-in-the-world provides a directional sense toward the environment and the intrinsic possibility of manipulating and transforming it. Such manipulation initially aims at extending existence, since every form of life strives foremost for self-preservation. The body embodies this impulse: adherence to life is not the result of deliberation but of our being thrown into and embodied within the world. Our bond with technology accompanies us from the outset: we are a technical animal and technical practices must therefore be sustained by a system of proto-values rooted in our embodied relation to an environment that is never neutral. In this sense, things "speak" to us, as Merleau-Ponty shows through his analyses of painting.

The critical examination of illusion reveals it as a structural component of subjectivity. Every form of empirical phenomenological appearing is, ontologically, a *virtuality* of vertical Being. What we designate as "illusory" depends on an epistemological stance, often involving the privileging of certain aspects of the visible over others. Our field of existence rests on a relative stability, a *metastability* of meaning embodied in the appearing of things. Possessing self-awareness, our form of life can move beyond mere survival to reflect on meaning, on being-in-the-world, and thus give voice to the *anguished wonder* (*thaumazein*) from which philosophy originates. The illusion of meaning permeates the structure of appearing: it motivates the need to justify the world so as to stabilize the sense and praxis of our dwelling within it. The analysis of illusion ultimately points to our inscription in the world through a blind spot that can never itself become an object of experience. Our body participates in the world as a complex organism, understood as an internal relation of reality.

If there is appearing, there is world, and therefore meaning (plane of immanence). The metastability of our form of life is sustained by generative mechanisms of meaning that belong to the plane of immanence and cannot be transcended. Ontological analysis of the real, pursued by the sciences, reveals instead that reality in itself bears no meaning except by reintroducing subjectivity. Illusions—from free will in the strong sense to the structuring of the perceptual field—accompany conscious life. Nietzsche already recognized art as the privileged expression of this need: art generates meaning by *instituting* a world. Artistic creation is not arbitrary; it seeks to express the generative process itself. The art of each epoch articulates its own paradigm of meaning, that theological-metaphysical background which has accompanied the culture of peoples and their works. Contemporary art, in turn, expresses the logic of technology: just as technology exposes processes, it renders meaning reproducible and progressively strips away the mystery of creation.

Art, religion, and philosophy are human prerogatives: our need for truth, for giving meaning (philosophy) to knowledge (science), for transcending the visible (religion and theology), our need to express feelings and emotions (art) confer meaning uniquely on our existence. No other form of life is structured by this order of meaning tied to such a degree of reflexive awareness.

The development of science and technology increasingly integrates us into a digital rather than a purely natural environment. Since our behavior is structured and therefore computable, it can be incorporated into broader techno-scientific apparatuses that we ourselves help sustain. Our alliance with technology, grounded in the extension of existence, is not accidental; yet processes of hybridization will inevitably reach a point at which they reshape subjectivity itself, producing a human being different from *Homo sapiens*. Such transformations are difficult, perhaps impossible to predict, given the non-linear pace of technological change. Consider, for instance, that the vast majority of data available to us has been generated only in recent decades. The conceptual categories through which we understand our form of life will require profound revision, and phenomenology will be called upon to rethink emerging forms of hybrid subjectivity. The decisive question remains: even as human beings transform themselves while retaining a high degree of self-awareness, will a transcendental horizon of meaning always be required to stabilize existence?

How will the unveiling of the mechanisms of subjectivity alter our ways of dwelling? Will technological rationality erode religious frameworks? Will individuals (whether secular, atheist, liberal, or progressive) who are deprived of shared values find new forms of social cohesion? Advanced capitalism appears to respond to a similar logic, absorbing contradictions and transforming desire into a resource within an adaptive apparatus. Through the reversibility of meaning, human action becomes integrated into a system capable of computation, prediction, and self-regulation. In this way, every gesture feeds the apparatus, and subjectivity itself risks becoming a functional component within a self-sustaining system whose ultimate aim is its own preservation.

Such a system tends toward an indefinite increase of its own power, seemingly without intrinsic limits. In principle, one cannot exclude that, in the future, a self-aware artificial form of life might emerge from it. For now, this remains within the domain of science fiction; nevertheless, let us allow ourselves a thought experiment. On the one hand, imagine the progressive and indefinite enhancement of human subjectivity; on the other, the possible emergence of self-conscious artificial life forms. One might even hypothesize artificial entities endowed with a lower degree of awareness, precisely in order to make them more adaptive, along the lines of viruses or insects. This raises a decisive question: must a living being, insofar as it is self-conscious, necessarily inhabit an order of meaning? The experience of a world is always intertwined, in a chiasmatic manner, with an order of meaning; this order functions as a necessary illusion of life. No matter how far technology may extend our will to power, our capacity for determination and manipulation, or the enhancement of our bodies, experience remains inseparable from a meaningful horizon. Even in the case of a being endowed with exceptionally high cognitive capacities, self-awareness would still be bound to an order of meaning—that is, to a *Gestalt*. One might further suppose that such a being, precisely because of its capacities, would also be aware that what orients and structures its existence does not, in itself, possess intrinsic meaning. Yet the more transparent this relational nucleus becomes, the weaker its effectiveness may prove in evolutionary terms. Why act in a given direction if reality, in itself, is devoid of meaning? The progressive unveiling of these dynamics (assuming such unveiling is possible) raises urgent questions concerning the stability of social organization. In this sense, the "essence of nihilism" is not an abstract doctrine but a structural possibility inscribed within the empirical and phenomenal appearing of the world.

Subjectivity is therefore always correlated with a world whose meaning orients the behavior of every form of life. Philosophy, understood as a rational response to the need for truth, stabilizes subjectivity; similarly, the interplay between the sacred and the profane institutes an order of meaning that fulfills an analogous function. Art, in turn, responds to the need to express the mystery of existence and of creation itself. In Western culture, religion, art, and philosophy have historically functioned as stabilizing structures of subjectivity, or, more fundamentally, as archetypal generators of meaning accompanying forms of life endowed with a high degree of awareness. If this is the case, then to speak of art, philosophy, and religion in a strictly post-human perspective risks becoming a contradiction in terms. Alternatively, one would have to acknowledge that these forms function as transcendental structures, conditions for the possibility of meaning rather than contingent cultural products.

As already suggested in the earlier thought experiment, a perfectly rational yet disembodied form of life (one not integrated into its environment through a body) would lack the conditions necessary for wanting to live. What would be missing is precisely the "irrationality" of the body: depth, affectivity, drives, the unconscious—in short, a bodily field that "thinks" prior to reflection and institutes a dynamic environment in which meaning can emerge. A future artificial intelligence endowed with strong self-awareness and perfect rationality would therefore require a body, understood as an integrated system embedded in its own environment and oriented, first and foremost, toward self-preservation. Only under these conditions could such a technical entity acquire genuinely biological traits, that is, those proper to living organisms.

IX.

Engaging with Merleau-Ponty today thus enables a critical reflection on the contemporary world and on the transformations driven by the acceleration of techno-scientific development, through a renewal of phenomenology understood as a philosophy of relations. This engagement does not exhaust the richness and complexity of Merleau-Ponty's thought, nor is that its aim; this book is not conceived as an introductory study of Merleau-Ponty, but as a problem-oriented inquiry that draws on his thought as a living philosophical resource. Rather, by focusing on perception as the epicenter of his philosophy, we have seen how a constellation of interconnected problems emerges.

The phenomenology of perception, grounded in an understanding of the body as both lived body and perceptual system, constitutes a privileged access point to philosophy itself. This initiation proceeds, on the one hand, through immediate experience and a critical engagement with Husserl, particularly his analyses of the mediations internal to experience. On the other hand, it opens a sustained dialogue with Gestalt psychology, whose relevance has been renewed by contemporary studies, and points toward further intersections with current research in psychology and neuroscience.

In Merleau-Ponty's spirit, these lines of inquiry can be developed in multiple directions, not merely through exegetical work but through new phenomenological analyses conducted in close dialogue with contemporary science. In particular, the analysis of the body as embodied subject opens up decisive questions concerning digital technologies and emerging forms of hybridization that increasingly affect subjectivity. Human beings have always manipulated their environment in order to adapt to it; today, this activity leads toward a "new flesh," that is, toward deeper integration with virtual and digital environments. To complete this integration, the body itself may need to be redesigned so as to maximize the exchange of information with digital worlds: after having adapted to the natural environment, it may now be called upon, metaphorically speaking, to shed its skin in order to integrate into a hybrid environment.

The scenarios unfolding before us mark the dawn of an era whose rhythm is dictated not by subjective time but by the autonomous acceleration of the techno-scientific apparatus. Technology no longer merely mediates our relation to the world; increasingly, it interacts directly with itself, generating new forms of knowledge through artificial intelligence. Without human mediation, development becomes exponential, while the risks and consequences of transformation remain largely unpredictable. We can, at best, discern the main vector animating this epochal shift: a progressive displacement of our relation to the natural world toward digitally constituted environments. Whether these transformations will still leave room for philosophical reflection remains an open question. It is unclear whether science will ultimately strip philosophy of its role, delegitimizing its modes of thought and the experiential ground that sustains them: namely, the capacity for astonishment and wonder before the appearing of visible and invisible phenomena. At this juncture, the gesture of *epoché* re-emerges as indispensable: a suspension of judgment regarding the ultimate outcome of these scenarios. For now, we can only glimpse their direction, the folds of a future that is already taking shape.

The distant echo that has guided this inquiry lies in Merleau-Ponty's lexical experimentation, which reactivates historical sedimentations while opening unprecedented conceptual paths. Do such choices weaken rational rigor, rendering them historically "obsolete"? Or do they, on the contrary, point toward something that remains to be thought, their metaphorical density nourishing new horizons of meaning?

By conferring historicity upon circularity and reversibility, the forms of life in which we are inscribed are sustained, stratified, and renewed. History thus remains open to new emergences, to forms of life whose destiny may not coincide with that of philosophy as we know it. Or else the philosophical question may reassert itself with renewed intensity, rediscovering its origin. As Merleau-Ponty writes: "The philosopher speaks, but this is a weakness in him, and an inexplicable weakness: he should keep silent, coincide in silence, and rejoin in Being a philosophy that is there ready-made. But yet everything comes to pass as though he wished to put into words a certain silence he hearkens to within himself. His entire 'work' is this absurd effort. He wrote in order to state his contact with Being; he did not state it, and could not state it, since it is silence. Then he recommences…" (VI, 125).

www.ingramcontent.com/pod-product-compliance
Ingram Content Group UK Ltd.
Pitfield, Milton Keynes, MK11 3LW, UK
UKHW020936180926
475692UK00003B/72
9783032270207